13 Dreams Freud Never Had

The New Mind Science

J. Allan Hobson, M.D.

Pi Press
New York

Pi Press

An imprint of Pearson Education, Inc.
1185 Avenue of the Americas, New York, New York 100036

Pi Press offers discounts for bulk purchases. For more information, please contact U.S. Corporate and Government Sales, 1-800-382-3419, corpsales@pearsontechgroup.com. For sales outside the U.S., please contact International Sales at international@pearsoned.com.

Printed in the United States of America

First Printing
10 9 8 7 6 5 4 3 2 1

Library of Congress Cataloging-in Publication Data:
2004110745

Pi Press books are listed at www.pipress.net.

ISBN 0-13-147225-9

Pearson Education LTD.
Pearson Education Australia PTY, Limited.
Pearson Education Singapore, Pte. Ltd.
Pearson Education North Asia, Ltd.
Pearson Education Canada, Ltd.
Pearson Educatión de Mexico, S.A. de C.V.
Pearson Education—Japan
Pearson Education Malaysia, Pte. Ltd.

Contents

Author's Note

Sigmund Freud correctly believed that dreams are a key to understanding the human mind. He was also correct in assuming that any scientific psychology needed to be brain-based. But lacking that base, he was forced to speculate, and I have found that his contribution to a science of the mind is, at best, obsolete and, at worst, misleading. My admiration for his ambition makes me imagine what he might say in the light of modern dream science.

This book begins with a mischievous daydream of mine about Freud, but it's constructed around 13 real dreams I had over the course of my career as a neuroscientist. Each of the 13 chapters begins with notes, sometimes sketches, too, jotted down as soon as possible after waking from the dream. I dreamt many of these dreams decades ago. Of particular interest to some will be the more recent dreams I had following my stroke. The stroke had intriguing affects on various neural faculties and showed up in my dreams in fascinating ways. Coupling the immediacy of those experiences with the knowledge I have gathered over the decades of my work on the brain, I hope to give readers a fresh understanding of what science has revealed about the mind over the last century since Freud bravely set out to bring science to the land of our imaginations.

Sigmund Freud Imagined

In 1885, when I was 29, I began to translate what I knew about the brain into psychology. Ten years later, in 1895, I was 39 and ready to write what I hoped would be my magnum opus, The Project for a Scientific Psychology. At the beginning of the 21st century, most people refer to it simply as "The Project." By basing my psychology on the firm foundation of brain science, I aimed to create a theory that was "perspicacious and free from doubt." It didn't turn out that way. However, I must say, brain science did progress in leaps and bounds and has created a new world for a real science of the mind.

The forces that interacted with my youthful ambition at the time I was hatching my Project were the result of the conviction shared by many scientists that materialism could finally triumph over idealism. At the Medical School in Vienna, my mentor was Theodor Meynert, one of the four original signers of the Pact Against Vitalism. Meynert and (through his influence) I were committed to explain all life processes, including the human mind, in terms of cells and molecules. For us, everything had a basis in physiology. Even our thoughts and feelings were bodily based, and our job was to show how this could be.

The Pact Against Vitalism was coauthored by the great physiologist Hermann von Helmholtz. It enjoined us from attributing any part of human experience to mysterious forces, in particular to vague and ill-defined "life forces" that explained away problems rather than accounting for them. All of this heady talk and our subscription to the pact were designed to keep us from falling into the trap of mystification that ensnared so many of our colleagues. I am proud to have been a part of this school of thought and insisted, to the end of my days, that physiology and chemistry would ultimately replace all my speculations about the mind.

One of the reasons I so much admired the English scientist Charles Darwin was that his *Origin of Species* flew in the face of creationism, a particularly pernicious form of vitalism. That I became an atheist at an early age is not surprising given the institutionally sanctioned opposition to organized religion and its commitment to mystification. By 1927, when I wrote my confrontational tract *The Future of an Illusion*, I had founded psychoanalysis and hoped it would be the new science of the mind. I applied it to religion and argued that all religious behavior and ideas were neurotically engendered. I even went so far as to suggest that religious faith was delusional—hence, not just superficially neurotic, but deeply psychotic.

Neurobiology appealed to me enormously. In the laboratory at Vienna University we worked on the claw opener muscle of the crayfish. It seemed a long way from crayfish claws to human thoughts and feelings, but we knew we were analyzing the movements of the lowly crab in terms of cells and molecules rather than attributing them to some mysterious life force. If movement could be so easily understood, it seemed clear to us that other dynamic life processes would also yield to neurobiological explanation. These are the things we talked about in the laboratory and in the Vienna cafes over coffee and cigars.

If I had been offered a full-time university position, I might well have taken it. Unfortunately, there were no positions available to me to pursue experimental science. As a young husband and father, I needed to earn my living by seeing patients. But I never abandoned my academic ambition. I was a well-trained neurologist and, by 1888, I was well-enough known and respected in academic circles to be invited to write an encyclopedia entry for aphasia, the impairment of language function that often occurs in stroke damage to the left upper brain. Aphasia, and other losses of higher human mental functions following brain damage, fit perfectly into my schema of a brain basis for all of psychology. Desires, drives, and even dreams were to be understood as brain-based processes. It is sad that we knew so little about the brain.

At the very time I was just beginning to see neurological patients, Charles Sherrington, working in England, was enunciating his famous reflex doctrine. That the motor system, including the claw opener mechanism of the crayfish, was governed by reflexive dynamics was obvious and helpful, as far as it went. But could I create a psychology based solely upon the reflex doctrine? I hoped so. And I tried. And I failed.

I knew that brains, like all animal tissues, were made up of cells. Our great anatomist-pathologist Rudolf Virchow had already formulated his cell doctrine: *omnia cellula ex cellula,* meaning every cell from a cell—implying, of course, a genetic mechanism for the establishment of the first cell and for the subsequent division of offspring cells during growth and development. Cells were recognized as the building blocks of the body's organs, including the brain and spinal cord. Sherrington used cell theory to account for reflex action in terms of competition between excitation and inhibition. But we knew very little about how nerve cells were constructed and even less about how they worked.

I was naturally drawn to the reticular theory of the brain put forward by the great German anatomists Wilhelm His and Rudolf Kölliker, who battled the Spanish upstart Santiago Ramon y Cajal. Cajal held that the nervous system was not a network of cells united as a syncytium, as the so-called reticularists believed. For Cajal, the cells of the brain really were connected as networks, but each nerve cell was discrete, bounded by its own membrane. Cajal's idea won the day, and his neurone doctrine led to many important discoveries that would have proven useful to me if I had stayed the course set by my ill-fated Project. But I really couldn't see what difference this heated dispute about neurones meant to my own brain-based models of the mind. It was simply too early for me to do what I wanted to do.

In 1895, I was 39 and in crisis both intellectually and personally. I was seeing neurological patients at home and often found no evident organic basis for many of their symptoms. Their complaints seemed very functional. At that time, I was a close friend of the otolaryngologist Wilhelm Fleiss. He believed that numbers determined behavior and that the size of one's nose correlated to sexual drives. Fleiss and I discussed many cases together. One of them, the famous Irma case, gave us professional grief when it was discovered that her persistent nasal symptoms were caused not by her birth date or the size of her nose but by a sponge left behind by Fleiss when he operated on her.

This medical negligence should have alerted me to jettison Fleiss sooner than I did. And it should have made me far more cautious in interpreting my dream of Irma's injection. But Fleiss and I were bonded by a growing conviction that many—if not most—of our patients' symptoms were expressions of suppressed sexuality. I would later refer to this process as repression, thereby indicating its unconscious origin. It is certainly true that sexual expression was quashed in late 19th century Vienna, especially in the women who were wealthy enough to afford our medical services.

The fact of the matter is that I fell for Fleiss and for many of his ideas even though I was too tough-minded to accept his numerological and nasal mystiques. He convinced me that the misfortunes and triumphs of life followed essentially astrological rules and could be analyzed and predicted using number theory. This was worst than vitalism. Yet I overlooked it. I was isolated and needed a colleague open to my own nascent ideas. I am appalled now to recall that I took his idea about the direct connection between the nose and the genitals at all seriously. When Fleiss operated on Irma, he thought he was curing her genital hang-ups by fixing her nose. Sheer madness.

But Fleiss was my friend and staunch supporter. As I grappled with my intellectual career, he was my ally. Later, when I had found my own way, it was easy for me to reject Carl Jung, my most important psychoanalytic coworker, because he believed in spirits. By then, Jung was also a thorn in my side because he thought I was overemphasizing the role of sexuality in the genesis of the normal and abnormal human psyche. Now that I think of it, I first rejected Fleiss because he embarrassed me with his numerological and nasal notions, but I never shed the obsession with sexuality we shared. When Jung challenged me on the role of repressed sexuality in the genesis of neurosis, I ostracized him on the grounds of his mysticism. According to my self-description, I was a rigorous scientist, not a necromancer. I knew that I was an adventurer and even called myself a conquistador. But I certainly never saw myself as a high priest in a cult of mentalism. This would mean I had fallen into the very trap I set out to avoid when I signed on to the pact against vitalism.

My Project for a Scientific Psychology, written in 1894 and 1895, failed so blatantly that I didn't even try to publish it. To me it was clear that I did not know enough about how the brain worked to create a satisfactory theory of the mind. I knew that without brain science I would be sailing in the uncharted waters of subjectivity, but there was no alternative, so I cut myself loose from brain science—or at least tried my best to do so.

Working hard seeing patients by day, I had time to work on my Project only in the evenings. It was a difficult task because I had so little scientific evidence to build on and so much of what I thought was solid science turned out to be wrong. For example, my concept of reflex action, borrowed, in part, from Sherrington, was flawed.

According to the neurobiology of 1895, all the energy—and information—that entered the brain came from the outside world. I never guessed that the nervous system could generate its own energy and information. Furthermore, I wrongly assumed that the system was unprotected from takeover by the invasion of external stimulus energy. To understand this idea, imagine a house that is unprotected by lightning rods when a thunderstorm breaks. Such a house is vulnerable to the heat (and fire) that lightning generates when it seeks a path to the ground.

Worse still, I assumed that once external energy and information entered the system they were obliged to stay there—building up a charge, as I supposed—until they were discharged as motor behavior. All these false beliefs, these delusions, were part and parcel of my Project. No wonder I couldn't get anywhere with it. And no wonder I gave up on that approach. I am proud to have recognized the futility of my Project.

Even as I turned away from the Project, the idea that the brain dealt with excess external energy by shunting it into circuits inaccessible to consciousness was still with me. I looked for a scientific outlet to unleash my ideas upon the world and have the same revolutionary impact I had hoped for with my Project. My ambition, which carried over from my thwarted academic career, was once again transcendent over my failure to use contemporary neurobiology to create a scientific psychology. Most people would have given up by this time. But not I!

As the clock was ticking down to my fortieth birthday, I was paying the rent—and other bills—by seeing more and more patients. In part through my friendship with Fleiss but mostly because it was just in the air, I became increasingly impressed with the notion that my patients, mostly single middle-aged women, were crippled by suppressed sexuality. Furthermore, these ladies were apparently relieved by talking about their problems with sympathetic doctors like me and Joseph Breuer, who was my own physician. Breuer shared his cases with me even if he did not follow my theories to the very end.

Another important event in my life was my trip to Paris in 1885 to work with the great French neurologist Jean Martin Charcôt. Charcôt's clinic was filled with the Parisian equivalent of our Viennese ladies. He had learned to use them to exhibit their neurologically impossible symptoms and to manipulate those symptoms via hypnotic suggestion. There was always something sexual in the background of these cases, according to Charcôt. His phrase "Toujours la chose genitale" stuck in my mind and there formed a complex with Fleiss's sexual theory to help me explain the observations made together with Breuer.

As the 19th century wound to a close, I was moving steadily toward the model of the mind that would make me famous: The unconscious mind is a cauldron of powerful and unacceptable impulses, many of them sexual, that constantly threaten to invade consciousness. In order to protect consciousness, these impulses needed to be pushed back (by repression), deflected (by displacement), or shunted to the body (somatized). This central notion of defense became the cornerstone of psychoanalysis, the intricate theory of the human mind that I built up and promoted so successfully in the first three decades of the 20th century.

My dream theory was constructed along the same lines as my theory of neuroses. Here it is in a nutshell: By day, our defenses against the invasion of consciousness by unconscious demons are generally strong. But the occasional inadequacy of these defenses is manifest in slips of the tongue and pen and in neurotic deformation. By night, our guard against unconscious impulses is brought down by sleep. To protect consciousness from overwhelming invasion, the mind resorts to the displacement, condensation, and symbolization that makes our dreams bizarre. This serves to protect sleep.

The more I thought about what to do next, the more I felt that dreaming was a good place to start my psychoanalytic theory-building. I said to myself that if I can't build a theory from the bottom up because I don't have the brain facts, I'll do it from the top down. I will deduce a dynamic psychology from the dream features, especially the apparent nonsense that some people call dream bizarreness. I will try to show that, like neurotic symptoms, they conceal hidden meanings that can be interpreted. An important concept that emerged from these considerations is that of infantile and even more primitive "wishes" that had to be banished from consciousness by active repression in order to avoid psychic catastrophe.

Just as slips revealed the volcanic unconscious impulses pressing toward the surface for unwanted expression in the daytime, dreams were for me evidence of a similar eruptive pressure at night in sleep. The secret of dreams, which was revealed to me on Pentecostal Sunday in 1896, was that the apparent nonsense of dreams was a function of the mind's effort to conceal the unconscious wishes welling up due to relaxation of the conscious mind as we sleep.

Looking back on this grand turning point in my life, and this epochal moment in Western thought, I am struck by two reciprocal ideas. One is that while I thought I was making a scientific discovery, I looked carefully at the experience of neither sleep nor dreams. The other is that by the forceful expression of my theory, I was able to persuade a remarkably large following of its truth. I was a good writer. A good speaker. Even charismatic, if I do say so myself.

However, I had brought most of the baggage from my abandoned Project for a Scientific Psychology into the dream theory despite my best efforts to throw it overboard. The reason that all this happened was because I was so determined and so convinced that I was right, that after I published *The Interpretation of Dreams* in 1900 there was no looking back.

It never occurred to me to make observations or collect data about sleep and dreams. Although most of the snippets discussed in my dream book were my own, I never kept a dream journal, nor did I ask either my colleagues or my patients to do so. I just knew, from my Pentecostal revelation, why dreams were strange. The mind needed to disguise and censor the instigating dream wishes to protect consciousness from disruptive invasion. Thus, dreams were the guardians of sleep. Once I had this key in my hand, I could unlock any dream door. Why bother to collect dream reports from people who could remember their dreams better than I could?

Take my oft-cited dream of Irma's injection:

A large hall—numerous guests, whom we were receiving. Among them was Irma. I at once took her to one side, as though to answer her letter and to reproach her for not having accepted my "solution" yet. I said to her, "If you still get pains, it's really only your fault." She replied, "If you only knew what pains I've got now in my throat and stomach and abdomen—it's choking me." I was alarmed and looked at her. She looked pale and puffy. I thought to myself that after all I must be missing some organic trouble. I took her to the window and looked down her throat, and she showed signs of recalcitrance, like women with dentures. I thought to myself that there was really no need for her to do that. She then opened her mouth properly, and on the right I found a big white patch; at another place I saw extensive whitish gray scabs upon some remarkable curly structures, which were evidently modeled on the turbinal bones of the nose. I at once called in Dr. M., and he repeated the examination and confirmed it…. Dr. M. looked quite different from usual; he was very pale, he walked with a limp, and his chin was clean-shaven…. My friend Otto was now standing beside her as well, and my friend Leopold was percussing her through her bodice and saying, "She has a dull area low down on the left." He also indicated that a portion of the skin on the left shoulder was infiltrated. (I noticed this, just as he did, in spite of her dress.)…M. said, "There's no doubt it's an infection, but no matter; dysentery will supervene, and the toxin will be eliminated."…We were directly aware, too, of the origin of the infection. Not long before, when she was feeling unwell, my friend Otto had given her an injection of a preparation of propyl, propyls…propionic acid…trimethylamin (and I saw before me the formula for this printed in heavy type)…. Injections of that sort ought not to be made so thoughtlessly…And probably the syringe had not been clean.

Naturally, I recognized the memory source of this dream. It mirrored my relationship to Fleiss and his misadventure with Irma, who was also my patient. But I failed utterly to recognize that Irma and her otorhinological condition were on my mind because of an anxious concern about malpractice. I did see that much of the thrust of this dream was an attempt to redress my error and put things right. This is certainly wish fulfillment. But I must confess that there was nothing unconscious about either my anxiety or my wish to fix Irma's nasal condition. And my dream censor was not working well enough to disguise this nasty bit of business, which I had learned to ignore or rationalize when I was awake. Honestly, practically nothing is deeply or genuinely psychoanalytic about my interpretation of the Irma dream. Why didn't I see this at the time?

Could it be that I had become a believer, a convert to my own ideas? Was I the inadvertent creator of a secular religion? My "discovery" was gratifying as long as I could maintain that I was a scientist and that my theory was scientific. When I said that I knew that one day all my ideas would be replaced by physics and chemistry, I really meant "confirmed" rather than "replaced." Of course, it never occurred to me that I might be dead wrong.

I could defend myself from the charge of pseudoscience by saying that the technology for studying sleep was not available in 1900. But when, in 1928, my psychiatric colleague Hans Berger described the EEG, I hardly noticed his discovery. I was by then 72 and pushing psychoanalytic theory into philosophy, psychology, and the social sciences. Ironically enough, 1928 was the year I published my attack on religion. My attack on religion in the previous year did not include any portend of this news and wouldn't have even if I'd published it a year later. I asserted that the future of the religious illusion was not bright, never thinking that my own ideology was essentially religious.

In 1933, I published my article "On Dreams," in which I attempted very significant revisions to my dream theory. But I did not mention Berger's discovery. And in 1936, when Loomis and Harvey showed cyclical changes in EEG activation during sleep, I was 80. By then I had such distracting problems to contend with, surviving the Nazis and cancer of the jaw, that it isn't surprising that I didn't see what was coming.

Instead, I steadfastly renounced every effort to neurologize my theory. I didn't want psychoanalysis to be taken over by neurology, since I had given up on the brain-based approach of my Project. The result was that my followers did not notice the slow but steady incremental growth in neuroscience that was gradually supplying the material needed for my Project in 1895. By 1950, the stage was set for a revolution I did not predict and frankly would not have enjoyed. I died in 1939, long before the house of cards I had so carefully constructed began tumbling down.

Chapter 1

Dreamstage with Elephants

REM Sleep and Creativity

Dreamstage is re-created on a stage. My friend, the artist Paul Earls, is adjusting his lasers front center. The TV image of the sleeper is on left wall L, where it intersects with Ted Spagna's photographic ceiling slides. The sleeper is in bed back right. I wonder if it will work—basically, if the sleeper will sleep.

Then I remember: It doesn't really matter if the sleeper sleeps.

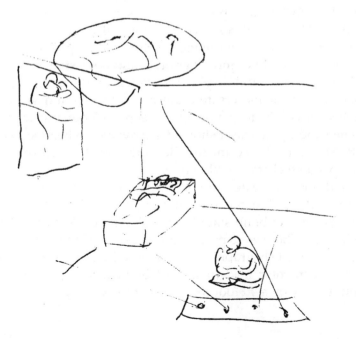

The scene shifts to an animated film about the brain. A whole brain is seen in 3/4 profile from above left.

Sections begin to be cut away much as if the brain were an apple.

These move away, rolling and tumbling.

When they near the edge of the table, they come to resemble elephants!

Then other sections detach themselves and roll away left.

These too turn into elephants. I marvel at the effect, which I take to be quite easy to reproduce.

Dreamstage, an experimental portrait of the sleeping brain, was an exhibit that first opened at Harvard University in Cambridge, Massachusetts in 1977. It toured nationally from 1978 to 1980 and became Dreamscreen in Bordeaux, France in 1982. The exhibit featured a person sleeping naturally in a soundproof chamber in full view of the visiting public. By recording the sleeper's EEG and other brain and body signals, the media artist Paul Earls could display the signals on the gallery walls via laser projections and create brain music via a synthesizer. When the sleeper was in deep, non-REM sleep, the brain music was low-pitched and a green laser showed slow waves; when the sleep was in REM, the music became fife-like with whistles corresponding to side-to-side flickers of a blue laser.

Ted Spagna created time-lapse photographic studies of sleep that revealed a dynamic sequence of posture shifts that coincided with changes in brain state from wake to non-REM to REM and back again. These were projected in lap-dissolve mode on the ceiling. The Swedish graphic artist Ragnhild Karlstrom composed three screen color field images of brain histology that changed continuously on a 40-foot-long wall of the gallery.

The "dark space" of the exhibit, where all these media were displayed, was carpeted and pillowed so that the shoeless visitors could sit or lie down to absorb the imagery. For the Bordeaux show in 1982, I added animated dream drawings and neuronal firing patterns to the dark-space media. In the "light space," the visitors could retrieve their shoes, read storyboard explanations of the science, and see drawings and photographs of the brain and of sleep behavior.

I recorded my Dreamstage dream, which I refer to as "Dreamstage with Elephants," on a Friday morning in the spring of 1982 while sitting in Boston's Logan International Airport, awaiting a flight to Paris. By then, I had discovered that by taking my journal with me on all trips, airport delays could become much more satisfying experiences. High winds raged outside, and my flight was being delayed by at least an hour. I knew I would miss my Charles de Gaulle connection to Bordeaux, where I was to continue my collaboration with French scientists and artists on a new exhibition to be called Dreamscreen. It offered an opportunity to present the imagery of the brain and sleep in a radically new way, and my mind had been working on it as if in a studio. The dream report quoted previously has a summary feel to it, as if important details were already being lost, but I had awakened and immediately committed enough of its very striking content to memory to be able to annotate and draw some of the dream's most interesting formal features.

The period from 1977 to 1984 was one of great ferment for me, my scientific coworkers, and the artists Paul Earls and Ted Spagna. One of our central interests was the autocreative nature of dreaming, and this dream serves as a good doorway into what interested us. Dreamstage and its spin-off projects were not terribly great art in every beholder's eye, but dreaming is fundamentally artistic in its efforts to recombine elements in aesthetically pleasing and original ways, so our marrying of art and neuroscience seems appropriate to me even now.

There is an almost tableau-like quality to the first scene of this dream. The set is hallucinated with many of the familiar Dreamstage elements visualized in a lifelike way. Nothing much happens. I don't move, and Paul Earls doesn't move, as would be likely in a REM sleep dream scenario. My thought processes are more rational than usual in my dreams, but I am nonetheless unable to reason incisively. Emotion is weak, but some anxiety about success is present.

In the second scene, the action picks up a bit, but the imagery is still very limited. I am obviously involved in envisaging new media to help make my points about dreaming and the brain more clear. But what seem like brilliant ideas during the dream turn out to be quite impractical when I wake up. It is only after I wake up that I have an idea that is both feasible and appealing. It is to photograph the brain to show the variety of its surface aspects and mix the images with shots of other natural objects as a way of calling attention to such similarities as these:

cerebellar sections ≈ cedar trees

cerebellar surface ≈ cauliflower

cerebral surface ≈ mushrooms

ventricular surface ≈ sea urchins

By changing the scale and using macrophotography to obscure the context of the flora and fauna, viewers would be unsure what was human tissue and what was not. It might make witnessing the seat of consciousness more appealing.

Indeed, the cerebellum has such a plant-like look that its branching lobes have been called arbor vitae (those cedar trees marking property lines across suburban America). With deep aesthetic pleasure, I have traced microelectrode tracks through this gorgeous cerebral structure. Colleagues and I at the Massachusetts Mental Health Center had to go through the cerebellum to get to the brain stem, where we had good reason to believe we would find neurones of great interest to theories of sleep. This turned out to be true, but because it was so difficult to record in the brain stem, we spent a lot of time in the cerebellum working on the Purkinje cells. These cells transform data about the body's position in space into repositioning commands executed by the brain stem. In the branches of the arbor vitae labeled lobules IX, X, and XI, we found many Purkinje cells that increased their firing rate in conjunction with the eye movements of sleep—hence the rapid eye movement, or REM, that everyone now associates with intense brain activation in sleep.

I regularly used my neurosurgical skills to implant microelectrode cylinders in the skulls of our experimental animals. Beholding the cerebellum directly, I was often struck by the resemblance of the cerebellar surface, which I beheld directly, to the cauliflower, one vegetable that I, like many children I've known, never learned to enjoy eating but loved to look at.

I had the same aesthetic twinge when I removed the brains of our cats to prepare them for histological sectioning. We used a combination of Nissl Stain, an aniline dye that colors neuronal cell bodies a light blue, and Luxol fast blue, which colors their fiber (or axons) turquoise. For Dreamstage and Dreamscreen, Karlstrom made a three-slide-wide color field display of these and other neurones.

Here, we said, are the business elements of the brain—the neurones. By then, we knew them to be active in sleep. And their activity was continuous, as if the brain were still processing data about position and movement even as we lie unconscious, apparently oblivious, and outwardly immobile in our beds. When our cats entered REM sleep, moved their eyes, and twitched, the brain cauliflower—usually bland—became truly *picante,* as if the brain were processing movement data that cannot be outwardly appreciated but is recognized inwardly as dream movement. Many of the dream animations I showed in Bordeaux were attempts by the dreamer to depict the curious movement trajectories of his dreams.

If the cerebellum (which the Italians call the *cerveletto,* meaning little brain) looks like a cedar tree, the cerebral cortex (which is much bigger and less finely convoluted) looks like a field mushroom. The cortical gyri form hillocks that increase the surface area. These hillocks are wider than those of the cerebellum because there are so many more interconnections between both local and distant neighbors.

I could see these glistening hillocks when I opened the skulls of living animals to implant electrodes specially designed to pick up the internal signals shot from one part of the brain to another during REM sleep. The mushroom-like cortical surface that interested me the most was the occipital (meaning posterior) or visual (meaning excited by light in waking) cortex.

Michel Jouvet had been my host and mentor during the training year I spent in Lyon, France in 1963. I knew from Jouvet's work that very large EEG waves could be recorded from the cortex during REM, but only if one of the electrode tips lay below the cortical surface in the fiber tracts (called the white matter). This was presumably because these REM sleep signals were directed to the cortex from the lower brain in pathways that ran perpendicular to the brain surface. Because they could also be recorded in the geniculate nucleus of the thalamus and the pontine nuclei of the brain stem, these signals became known as PGO waves. The relevance of these internally generated signals to dream vision was immediately suggested. The PGO waves are still the best candidates we have for the role of dream stimulus.

In Dreamstage (1977–1982), we could not easily incorporate all these important neurophysiological details because we were stressing sleep and dreams in humans. But by the time I was going to Bordeaux, and had this dream, I was trying to do so. I used white-on-black negatives of our oscilloscope tracings to show this activity via the rapid slide change technology afforded by computerization of projectors.

A pedagogical note: The danger of trivializing the complexity of brain parts with plant life was more than counterbalanced by the representation of what goes on under the surface. But that the forms of the surface are homologous to familiar objects such as vegetables makes two important points at once. The first is that nature is economical, using similar forms for different purposes. And second is that we can overcome our distaste for brains by recognizing their similarities to familiar objects.

When visitors to the Dreamstage light space looked at David Scott's 3-D electron microscopic photographs of the brain's ventricular surfaces, it was impossible not to think about the form of tidal pools seen on rocky coasts. The ventricles, which are fluid-filled cavities around and within the brain, appear smooth to the naked eye. But the scanning electron microscope shatters that illusion, revealing instead a dense and variegated array of strange plants and bizarre animal forms. There are the cilia, which beat to keep the cerebrospinal fluid moving, and there are large tentacular creatures, looking for all the world like extraterrestrials, that might function to keep the surface clean, like an automatic swimming pool vacuum.

What does all this have to do with sleep and dreaming? Nothing that we know for certain now, but it seems unlikely that this fluid-filled system functions only to cushion the brain against shock by keeping it afloat. The brain's canals, lakes, and rivers are available for transport too. So even if cushioning is a primary function of the fluid, we would be surprised if crucial molecules were not traveling up and down those cerebral waterways. We know, for instance, that many serotonin-producing neurones of the brain stem send axons toward the ventricular surface, where they end. Is serotonin secreted into the ventricles? We know that fluid from the brain stem, which contains the NREM-REM sleep oscillator, has access to the ventricular surface of the hypothalamus (which my Harvard colleague Cliff Saper has shown to contain the sleep/wake switch). These structures communicate with one another more rapidly and reliably by projection fibers than they could by fluid transport. The slower, smoother fluid route might serve to round off the rough edges of abrupt state transitions— such as waking up or falling asleep—that seem to take minutes (or even hours) for us to accomplish.

I nod to Freud and his relentless focus on sexuality in relating my weirdly salacious idea to match brain images with ambiguous photo images of the body surface, as follows:

Cortical gyri	Lips, fingers
Colliculi	Thighs, buttocks
Cerebral aqueduct	Navel
Pituitary	Penis

This analog has been on my mind for some time. Whether my Dreamstage dream has anything to do with my imagining them on my way to Bordeaux remains uncertain. Nevertheless, the photographic ambiguity I am contemplating might depict many of the surface similarities that I have noticed as I work on the brain.

The cortical gyri look enough like lips and fingers to create visual confusion by breaking contextual and scalar rules. Fingers and especially lips are sensual structures, a feature not usually associated with the brain. Yet while the brain is the seat of all sensuality, its surface structure is also sensual. Some may find this gratuitous, but the experience is real enough. In any case, my goal is to share the experience of scientific discovery with the people who have paid for it all, not exclusively my parsimonious peers.

The colliculi are particularly seductive hummocks that sit like thighs or buttocks on a hill above the midbrain. I used to marvel at them when, having removed the anterior lobe of the cerebellum, I could direct instruments into the brain stem using the opening of the cerebral aqueduct as a landmark. They were so plump, so beautifully rotund, and so well lubricated by their own juice, I wanted to say, "Here are the colliculi! Aren't they wonderful looking?"

The colliculi represent nature's first go at creating a cerebral cortex where signals from the outside world could be processed. In this area, auditory signals to the inferior and visual signals to the superior colliculi are evaluated in the service of redirecting attention. So, return your attention to this page and stop being distracted by the birds outside the window—do the bidding of the colliculi, the little hills of the brain.

The opening of the cerebral aqueduct is the gate of the canal linking the third and fourth ventricles of the brain stem. It sits there, pouting, like puckered lips, and like some people's belly buttons, all obscure objects of desire if we follow Bunuel closely. Portals all, between one part of the self and the other.

The pituitary glands at the stalk of the cerebral aqueduct make up a structure that is recognizably phallic. It should not, therefore, be surprising to learn that this structure is, among many other important things, the producer and distributor of molecules that are crucial to growing body parts that are necessary for successful reproduction. Follicle stimulating hormone (FSH) and luteinimizing hormone (LH) are essential to the development of secondary sex characteristics. They are both produced in the hypothalamus and stored for delivery in the pituitary gland. And that delivery occurs in sleep! So does the delivery of growth hormone by which our body mass and its ratio of muscle to fat is determined.

The isomorphism between pituitary and phallic structures thus is profoundly functional as well as structural. And the strict state dependence of hormone release is an unexpected bonus to this analogy. Growth and sex hormone release are 90–95 percent sleep-related. It is no coincidence that adolescents, growing at extraordinary rates in a maelstrom of hormonal activity, sleep so annoyingly long and deep.

Dreaming is creative. To the extent that dreaming does not replay whole memories, it stitches together fragments and fills in with newly synthesized mental products. Brain slices that become elephants might not be exemplars from an aesthetic point of view, but they are creative. And throughout human history, people have been waking up and making art from their dreams.

At the time of my dream, I had become so absorbed in Dreamstage that I found it difficult to do laboratory science. It wasn't just the extensive traveling; it was the constant cognitive and emotional stimulation of struggling with a new art form. As Federico Fellini immediately recognized, Dreamstage was closer to theater, with all the energy of thespians putting on a show, than any other art form. I was an amateur in this domain, but the tremendous popularity of the topic and the pleasure of playing in this new medium were with me constantly.

The exhibit was never, in fact, on a stage. There was no true stage at the Carpenter Center, where Dreamstage ran for two months in 1977, and there was no stage in its U.S. traveling phase when it visited six cities from 1978 to 1980. Putting it on a stage in my dream is a sign of the literalness and concreteness of dream thought. I had coined the word *Dreamstage* to suggest a process more than a physical setting. For me it was the stage—or phase—of mental activity where imagination ran wild. And my imagination certainly did run wild in my dream that night.

Paul Earls was not a part of the Bordeaux installation because neither the organizers nor I felt that the laser projections of the sleeper's EEG, EMG, EOG, and EKG were scientifically satisfactory. We knew we could project physiological signals directly as part of the giant video image of the sleeper, and that is what we did in Bordeaux. But my dreaming mind didn't take account of that change: It put Paul there, adjusting his lasers. Paul was my first Dreamstage collaborator and by far my most active coworker in the planning and producing of the Carpenter Center exhibit. So his presence in the dream is emotionally salient even if it is out of date. In Bordeaux, I was much more on my own. Via images and words, I wanted to explore new ways of telling my story.

Ted Spagna's time-lapse animations of the nights of sleep of his family and friends were always fascinating and entertaining. And although Ted didn't go to Bordeaux either, I took his slides with me. They had the same powerful effect on the French visitors as they had on Americans. The technique of time-lapse photography had been accessible since the work of Edward Muybridge and Etienne-Jules Marey in the late 19th century, but Spagna was the first to use it extensively and successfully to photograph sleep. In my dream, the overlap of Ted's Circular Projection Screen and the video image of the sleeper was a device of my dream devising. Psychoanalysts might call this a condensation and assign a defensive function to it, but it is more simply seen as a purely cognitive condensation. My mind saw these two media as superimposable, and they certainly are—even if we never superimposed them in the reality of the exhibit. My dreaming brain-mind has become quite concrete in its linking of associatively related images.

Putting the sleeper in a bed in the back corner of the stage is also a dream choice that the waking mind would immediately reject. We always knew that the sleeper had to be central to whatever exhibit design we chose, because we wanted no possible misinterpretation of the human source of the live data. In Cambridge, the sleeper's visible room was in the center of a wall; in the U.S. traveling show, it was a two-window lozenge in the center of a circular space. In Dreamscreen, it was a cube in the center of the space, but its large roof did serve as a "stage" for the video and other projection equipment. This included animated dream drawings—and their synchronized texts (translated into French). That marked the first time we had actually introduced dream content into the exhibit. It worked beautifully. My dream omitted this important modification.

Just because I wanted Dreamstage to be theatrical doesn't mean I was unconcerned about the scientific integrity of the content and the didactic efficacy of the media. The central messages that I hoped to get across were two: that in sleep the brain was internally activated, and by understanding the brain's activation in sleep we would learn more about dreaming.

In April 1977, when Dreamstage opened at the Carpenter Center at Harvard, Robert McCarley and I had already published our two back-to-back articles in *Science* about the brain stem's role in sleep cycle generation, and our two articles on the brain-based dream theory were about to be published as lead pieces in two successive issues of *The American Journal of Psychiatry*. We had worked hard. We had come up with novel findings and proposed new models. I wanted the world to notice. But I didn't want to confuse the world as I tried to educate and promote brain science.

From beginning to end, the Dreamstage exhibit at Harvard included a bona fide sleep lab. The center of the exhibit—and our new science—was a real person, sleeping in a real bed, in real time. The sleep had to be natural, so no sedatives or drugs of any kind were used. If the sleepers followed my rule not to sleep anywhere else, they all were able to sleep in the exhibit chamber with hundreds of people watching them.

But it wasn't all smooth sailing. When the father of our Carpenter Center sleeper read about the show in *The New York Times*, he correctly guessed that it was his son who was the star of a show that, for various reasons, was open in the daytime. Since he was paying MIT a hefty tuition fee, he was so vexed about his son missing classes for a month that he came to Boston unannounced to confront me and our star directly. Fortunately, our sleeper was no longer a minor and thus was free to work for us if he saw fit.

When we opened our road version of Dreamstage in San Francisco's Exploratorium a year later, I had a more severe scare when our West Coast sleeper, whom I will call Janice, told me she thought we were sending her mind-altering messages through the EEG waves! Janice was a graduate student in psychology at Stanford. She had taken my injunction against sleeping elsewhere so seriously that she had sleep-deprived herself into a paranoid psychosis. This was alarming in itself, but when Janice told me that she had misconstrued her unfastened seat-belt buzzer for a pursuing state trooper, I advised her to go home and get some sleep.

But sleep in the exhibit all those young Dreamstage sleepers did. For visitors who wanted only an impressionistic image of sleep, Paul Earl's synthesizer music and laser projections were driven—online—by the sleeper's brain waves so that the deep, non-REM sleep periods that occur early in a sleep bout could be seen as high-amplitude green laser wobbles on the wall and heard as basso rumbles in the music.

And every 90 to 100 minutes, like the clockwork that it is, the brain activation of REM sleep could be perceived by a shift to lower amplitudes in the green laser trace and a shift to higher-pitch registers in the music. To these signs, we added a blue laser plot of eye position that zigged and zagged with the REMs and triggered a whistle from the synthesizer. Anyone, even a psychoanalyst, could tell what sleep stage was being broadcast simply by looking and listening. We didn't have to label anything.

For those who wanted technical detail, we provided it by means of standard EEG brain waves, eye movement, and muscle potential tracings on an ink-writing polygraph. Our open sleep lab came complete with a trained technician who demonstrated the data and explained the niceties of scoring sleep records. Most people preferred to lie down on the soft floor, lay their head on a wedge-shaped pillow, and let the exhibit itself do the talking.

Outside this dark space of sleep and brain phenomena, explanatory material also was presented in what we called the Light Space of Scientific Documentation. One theme of the light-space media was "The Scientist-as-Artist," in which I emphasized the aesthetic nature of many of my colleagues' observations. One of the most exciting of the scientific storyboards was made by my Dreamstage collaboration with the photographer, Ted Spagna. His photos of sleepers made with the boom-mounted Zeiss Ikon—a time-lapse rig he had put together—revealed a clear-cut pattern of posture change during sleep. People were likely to toss and turn only if they were having trouble falling asleep or when their brains were switching into and out of REM after they fell asleep.

We confirmed these observations and their organizing hypothesis at the Boston Museum of Science installation of 1980, when we were able to study the sleep of several subjects recruited from the cadre of hospital night workers who sleep in the daytime and are comfortable with technical instruments such as a polygraph.

Twenty years later we now can study subjects sleeping at home using the Nightcap portable staging system that grew out of Dreamstage. We added a piezoelectric bandage that the subject applies to his eyelid. Along with the forehead-based accelerometer, which is sensitive to trunk rotation, we can now diagnose wake, NREM, and REM sleep with sufficient accuracy to conduct field studies of the cost and multiples of the yield of standard sleep lab studies.

Dreamstage thus proved to be much more than the publicity stunt it set out to be. As with my teaching and writing, I learned that changing the context of work and the language used to express concepts, data, and theories changes the methods and scope of science as it evolves. For example, it became clear that via the reach of the Nightcap we could study the mental states of our subjects around the clock. This meant we were studying not only the states of sleep but also the states of waking. And this meant we were after much more than a mere brain-based theory of dreaming, but a brain-based theory of all conscious states. The study of sleep and dreams is only an integral part of the study of consciousness. What is consciousness? How is it made? And what is it for?

At the Harvard installation of Dreamstage, we had to enclose the sleeper. It was difficult to create conditions favorable to sleep inside our sound-attenuated enclosures. We never tried setting the sleeper out in the open as in my dream. And this dream percept makes me worry. Will the sleeper sleep? The anxiety is immediately dismissed by my recall of the deliberate determination not to care if the sleeper slept in Dreamstage. We just couldn't afford to worry about that, and we told the sleepers not to worry either. Their job was to stay awake while not in the exhibit. If they did that job well, I was sure that sleep would take care of itself. And it did. So this dream is clearly about the new exhibit, but it is not brilliant in suggesting any new configuration. If anything, it has lost ground, because the exhibit is represented so loosely and erroneously.

I was not consciously aware of these defects, so I cannot explain the scene shift to brain animation, but I do know that my mind was often playing with novel ways to exhibit the brain. At the time of this dream, I had already begun to assemble films made by other neuroscientists. One, by Bob Livingstone, showed a brain being sliced away in a speeded-up time-lapse sequence that enabled the viewer to walk through the brain from stem to stern.

But my dream brain is considerably more rambunctious than Bob Livingstone's film brain. It shows dream animation more than it shows filmic animation. The dream brain pieces have a life of their own! They detach themselves and roll away, turning into elephants as they go! In this respect they resemble Clinton Woolsey's animalculi, images of the body surface—in the brain—that I had included in the Dreamstage Light Space as a way of showing other scientists' aesthetic responses to the brain.

One of the easiest ways to understand the value of a formal analysis of dreams is to consider the discoveries we have made by focusing on dream movement. Of course, detailed content analysis could come up with the same findings, but it hasn't. We think that is because physical movement is considered commonplace if it is not overlooked entirely.

Our own sensitivity to the pervasiveness and exotic nature of dream movement was conditioned by finding activation of brain movement systems wherever we looked in the brain of REM-sleeping animals except for the final common path motorneurones. They were turned off by active inhibition. It was this gate preventing motor output that Michel Jouvet and his collaborator, François Michel, picked up as the complete loss of muscle tone during REM in early animal experiments. Were it not for this disconnection between the outside world and motor pattern generators of the upper brain, we would all get up and act out our dreams the way Michel Jouvet's cats did and the way some fellow humans do when this inhibitory motor gating mechanism is damaged.

We always imagine movement in our dreams. And we are generally at the center of the fictive reality. My Dreamstage dream begins with a first-person-centered vision of a three-dimensional dream space. In the second part, the animation is embedded in the objects I want to manipulate. Animation thus is the order of the dream as much as it is the order of the film.

Until I looked for movement descriptions in REM sleep dreams and found them everywhere, I greatly underestimated their prevalence. I have therefore changed my language to emphasize visuomotor imagery rather than merely visual imagery. The fact that upper brain motor systems are activated in REM and the fact that there are REMs in REM sleep clinch this point and change forever our view of dreams as mere replays of waking experience.

The new view of dreams as the conscious experience of integrated visuomotor activation in sleep helps us understand not only the mechanisms of onirogenesis but also the developmental and plasticity hypotheses of REM sleep function. We are a long way from Freud's notion of the brain-mind as a reflex engine doomed to repeatedly shrink in horror from repressed wishes. Our REM sleeping brains and our REM sleep dreaming minds show us, beyond the shadow of a doubt, that we are much more complicated, much more interesting, and much more resourceful than Freud's reflex model suggests.

While we are asleep—in our beds or in our mothers' bellies—our brain-minds are creating a fictive universe. This creativity makes us agents in understanding our worlds from a very early time in our developmental life. And we return to this fictive world for considerable amounts of time throughout our lives.

My dreaming mind is seduced by its own creativity. It is certainly bad judgment on my part to think that such an animation effect as my brain-slice elephants would be easy to produce in the real, waking world. Dream visualization is deceptive because it is so easy, so plastic, and so dynamic. When I wake up, I realize how difficult—or even impossible—it would be to achieve these special effects.

The point of such a project would be to show that brain-like structures are everywhere. Look around you and you will see how natural a creation the brain really is. And it is yours! This was the mission of Dreamstage. Sleep is everywhere. If you will just look at it, you will see its internal dynamism. Dreaming belongs to everyone. Just pay attention to your dreams, and you will marvel at your own creativity.

As is often the case when I am about to leave for France, my dream language begins to assume a Gallic flavor. In France, I announce that the Brain is *Not* a *Choux-fleur* (even though it looks like a cauliflower) and the Brain is Not a Piece of *Fromage* (even though it can be sliced like cheese). Once I arrive in France, I begin dreaming in French, but the subject matter still does not fit the local conditions at all well.

This dream attempts but fails to solve a design problem for the third manifestation of Dreamstage. On the eve of my departure for France to participate in a two-week planning session with my French collaborators, the dream contains none of the innovative media that were developed in Bordeaux. In fact, most of these innovations had already been conceived but did not appear in the dream at all! So my dream is out of date. So much for the practical creative power of dreams with respect to real-life problems. But the failure to affect reality does not mean that dreaming is uncreative.

One novel feature of Dreamscreen in Bordeaux was the spatial setting, a 15-meter-on-a-side cube of space in the L'Entrepot Lainé, an 18th century wine warehouse that was placed at our disposal by Roger LaFosse, the artistic and administrative director of Sigma. A second was the addition of 1,500 high-quality neurophysiology images, hand-painted as reversed negative slides by Marie Tancrède in my lab in Boston. These images were made to dance on one of the five Dreamscreen screens in the space above the sleeper's cube by Gerard Lion's ingenious computer-driven projection display. The dream drawings were animated to show dream movement, and the corresponding dream reports were played through speakers in the pillows where visitors could lie, look, and listen to these fascinating reports.

This dream is evidence against the theory that dreams function to solve problems. Not only does the dream not solve the problems it poses, but it ignores solutions that have already been made. The brain-form slide show might be worth a try, but I had already thought of that before dreaming about it. My dream about it was no more creative and considerably less critical than my waking consciousness. As is often the case where dream function is concerned, negative examples such as this one usually are not put forward.

Fortunately, Dreamscreen had a much brighter future than was envisaged here in my dream! This example shows that dreaming often parallels life experience but is quite different from it. It has long been held that dreaming is a replay of memory. But this dream—and Freud's Irma dream, for that matter—shows that dreams represent only short memory fragments, not whole waking scenarios. Our systematic studies of dream memory indicate that less than 4 percent of dream content is actually a replay of waking experience. So what is going on?

What is going on is a matching of emotional salience and the myriad of real and imagined details requiring design resolution for the Bordeaux installation. In this case, my concern and apprehension about my performance as the director of Dreamstage are the emotions that drive the dream. My mind is indiscriminate and overinclusive in dredging up old ideas and spinning out new ones. But this very prolificacy guarantees the association of all these ideas with the emotional vector of my voyage. Far from disguising unacceptable impulses, this dream reveals the plenitude of possibilities that are tied to my hopes and dreams for Dreamstage.

Chapter 2

HMS Marathon
Turning on the Social Brain

Dean Tosteson is speaking to me, on Monday, about an upcoming event, but it's not clear that he wants me to attend. Later, on Friday, I go to a table near which I had spoken with him, and I find a note saying, "Allan, of course you're welcome to come to the Tuesday meeting," and giving particulars.

The note is scrawled across the top of a large, promotional-type paper bag on which is printed a color photo of the dean in full academic regalia, speaking at a lectern. In front of the lectern and, confused with Tosteson's image, is Dan Federman, his associate dean, whom the Harvard students call "the Fish."

Runners are milling about, awaiting the start of a marathon, to be run indoors. I am a confused, reluctant entry and, like the others, thin clad. I notice that my Harvard Medical School classmate, Kitty Beck, is standing next to me—an even more unlikely entry than I! We kneel (or rather squat) at the edge of the track in an intimate hunker.

I ask, "How long is it?"

"50 to 100 laps," replies Kitty.

I do some rapid (but vague) calculations and conclude that since the track is considerably less than the standard quarter mile in circumference, there is no way it is a full marathon distance race. But since the exact length is unclear—maybe 6 to 12 miles?—I am still uncertain about my ability to finish.

"What is your pace?" I ask. "Maybe we can run together."

"1/2 to 1 minute per lap," she replies, apparently intending (but still failing) to reassure me.

Still squatting, we kiss sweetly and innocently.

When we stand up, we are in the midst of the Hispanic women's contingent, and several olive-skinned contestants are also waiting to be kissed.

I feel suddenly diffident: It is too much of a good thing and utterly inappropriate.

So I begin to run, complete a couple of laps, and then, inexplicably, stop.

I woke up at 6:00 a.m. after what seemed to have been a long, exhausting dream about Harvard Medical School (HMS). As is often the case, I could recall the details of only fragments of this dream but was sure it had many other parts that eluded me. My estimate of its duration is 30 minutes, of which the two remembered fragments constitute no more than 5 minutes. If I am correct, I have lost 75 percent of this one. Sleep laboratory studies reveal that awakenings instituted in REM sleep yield detailed recall in 80 percent of the cases. This number rises to 95 percent if the awakening occurs during a cluster of REMs. If no awakenings are performed, recall may be nonexistent or fragmentary, as in my case. Most of our dreams are simply not recalled.

My Marathon dream was recorded as part of a project to compare dreams and fantasies. Each of the contributors of data was asked to record ten of each type of mental product for formal analysis. What we found was that dreams had more bizarreness of character identification than do fantasies, and this is certainly true of both fragments in this example. Dean Tosteson's face is confused with that of Dean Federman, a character incongruity that Martin Seligman has called dream transmogrification. Although it is not uncommon in psychosis, transmogrification *never* occurs in the waking fantasies of normal people. And I cannot imagine Kitty

Beck as a fellow long-distance runner—even in fantasy—although as a classmate and faculty colleague she might fit into that category metaphorically.

Because of my poor recall, normally I would not have recorded this dream. Yet looking back on it from the point of view of content analysis, I realize how loaded and charged even such poorly remembered dream fragments can be.

My professional life has been deeply embedded in Harvard Medical School. This dream springs from a set of related issues associated with my career that were on my mind at the time. My work with the William James Seminars was coming to an end. My relationship with the dean and his program was about to decline sharply. The death of Norm Geschwind, a friend and colleague at the school, was both a personal and professional loss with consequences for the security of my position within the school. After seven years of a tenuous political alliance, I was on my own again.

In the dream, I was struggling against all these odds to hold things together, and my brain was helping me synthesize an understanding at the same time its dreaming state was limiting my brain's analytic powers. In the dream, I did not recognize the impossibility of the fusion of Tosteson-Federman images. This image makes sense only at the deeper level of their political alliance. My competition with them was clear—transparent, as I like to say—and it was clear that on their turf I would lose. I was also unable to see that in my dream, time was as unconventionally structured as people were. How could it be Monday *and* Friday at the same instant and then refer to Tuesday? This temporal discontinuity is a cognitive defect that my emotional salience-seeking limbic system uses to its advantage.

This is the central point of my new dream theory: By ignoring certain errors of fact, the dreaming brain can consolidate and integrate emotionally salient content effectively and efficiently. The new theory turns Freud's distinction between primary and secondary process on its head in two important ways:

First, it says that yes, dream mentation *is* primary process, but it also says no, it is not disguised. Instead, dream meaning is strikingly evident.

Second, it says that yes, condensations, displacements, and symbolizations do occur in dreams, but these psychological processes are in the service of integration rather than serving disguise or censorship.

My dream motives are to curry favor (in order to accomplish shared intellectual goals) and to enjoy life (in the face of impossible assignments such as indoor marathons that only such citadels of ambition as Harvard Medical School could possibly sponsor). A few surreptitious kisses (however politically incorrect and personally misdirected) are a small fee to extract from such a demanding system. My motives are ambivalent, but they are not disguised by the dream; rather, they are revealed by the dream. They require only close attention, not complex interpretation, to be understood. And that is because their emotional salience is so clear.

None of this should be taken to suggest that I dreamed this dream in order to become aware of my motives. My motives were already known to me. If this dream's content is a clue to how my unconscious brain-mind operates in sleep, it says that I keep track of important relationships and events in my life via an emotionally driven shorthand. It is in this sense that I find common cause with Freud's dictum that dreams are the royal road to the unconscious. But my definition and concept of the unconscious are quite different from that of Freud.

Running, and then stopping, can be analyzed symbolically. I liked the new dean and at first worked hard to further his agenda. But then I stopped collaborating and went off on my own to found the William James Seminars. Dream movement and its cessation may also have a formal basis in the motor pattern generator activation of REM sleep. We know that the automatic brain stem instigators of walking and running are activated in REM. That could be why so many dreams, including this one, are animated.

The setting of the dream was noted to be "a kind of commencement–reunion–field day event with multiple, complex activities." I assumed it was HMS because of the personnel and plot actions, although there are many inconsistencies that defy that assumption. In fact, I had participated in the 1984 Class Day exercises, but that was in June, five months earlier. In my opinion, one stimulus for the dream was an article that had appeared in the Harvard Medical Alumni Bulletin the week before.

This article was important to me because it described my Class Day talk about my experience as founder and organizer of the William James Seminars at HMS, an activity that grew out of my early involvement with Dean Tosteson in his efforts to reform both the curriculum and the departmental structure of HMS. When I realized how much unpleasant political activity was involved in committee-based action, I had gone off, on my own, and initiated the seminars out of my two introductory courses in psychiatry (700a and 700b), which I had designed hoping they might be more scientific than their predecessors.

Some of the numerous tensions that resulted from my actions are clearly reflected in the dream.

First, I am not sure whether the dean really wants me to attend the meetings he mentions, because one of them has already taken place. But since this is *my* dream and not his, I have to take responsibility for the ambivalence expressed. This is classic displacement, but there is no need for it as a disguise because I am already consciously aware that I don't want to attend those meetings. It remains possible, however, that I don't want the dean to know that!

Second, I am discomfited by taking part in such a self-promoting event as this dream Class Day. This is made clear by the dean's note to me, which is written on a large paper bag. This bag is typical of those issued by organizers of corporate annual meetings and resource developers at educational institutions— both of which I abhor! And it is in the dean's photo on the bag that the transmogrification to Federman's fish face occurs. Self-promotion was exactly the charge leveled against me when Dreamstage had its huge popular success. And it was true: I was telling my own scientific story through the exhibit. So I must again share partial responsibility for the paper bag photos of the two deans! When, as the first appointment of Dean Tosteson, I achieved the full professor rank I had worked so hard for, the chairman of my review committee, Manfred Karnovsky, said, "We gave it to you in spite of Dreamstage!" This leads to my next point.

Third, I am ambivalent not just about Dean Tosteson, his program, and his style, but also about his administration! In the spring of 1977, when Dreamstage was still showing at the Carpenter Center for the Visual Arts on the Harvard campus in

Cambridge, I was greatly surprised to be invited to have breakfast with the then-Dean-to-be Tosteson. Why was he consulting me? Because his medical school roommate, Ed Evarts, had given him my name. When Dean Tosteson asked me if I would consider a position in his group, I told him I was hopelessly impatient with administrative work and that I wanted to devote myself to my research. But I added that I would work vigorously for the academic reforms he wished to promote and especially help foster interdisciplinary work in neurobiology, neurology, and psychiatry. Unfortunately, this very desirable integration across departments is still a work in progress.

What was the second dream stimulus? Not only had the Alumni Bulletin appeared the previous week, but also, on November 7—just six days before the dream—I had attended the funeral of Norman Geschwind. Norman, who had died suddenly of a heart attack, was one of the great minds at Harvard Medical School and important to me personally because he was a champion of mind-brain isomorphism and the sole advocate of the clinical sleep laboratory idea at Harvard. At his funeral, I had stood next to Deans Tosteson and Federman, united for the moment in mourning the extinction of one of our great lights. But this aspect of my life experience does not appear in the dream. Why not? Has it been repressed? I can't see why. It seems more likely that only snippets of life experience appear in dreams. A recent systematic study of narrative memory conducted by Roar Fosse and our group confirms this hypothesis. Only fragments of waking recollections are included in dreams.

My Marathon dream can be interpreted in a straightforward way without indulging in symbol decoding. Fragment 1 (Dean Tosteson) offers an interesting snapshot of my relationship with HMS and its leadership. I can see the connection between this dream and others concerning authority figures. I am nervous about them—quite naturally, I think. But does this dream tell me something I don't already know? I don't think so. In fact, I think my conscious memories of my life, even twenty years later, say much more about the important psychodynamic issues involved than the dream itself does.

Fragment 2 (runners are milling) is more bizarre, by far, and certainly invites speculation about symbolic meanings. Let us see if they really help and, if so, how they help. What is at stake here is not dream symbolism itself. All mental processes are symbolic. What is at stake is whether dream symbolism conceals covert motives that can be uncovered only by dream interpretation.

Our first problem is to understand the scene shift from what is at least somewhat like a medical school event to one that could not possibly be such despite the persistence of medical school themes and personnel.

Not only are marathons not standard Class Day events, they are never—as far as I know—run indoors. And neither I nor my Harvard Medical School classmate, Kitty Beck, would ever—even in 1984—consider competing in such an event. Nor would we be likely to be found in the "intimate hunker," a squat that I substitute for conventional race start posture. Does this mean that the race is a disguised symbol of something else? A hidden sexual motive, for example? Those committed to such formulae would say, "Of course; isn't it clear?" but I don't see it that way at all.

Kitty was a psychoanalyst who married my medical school classmate Anton Kris, the son of Ernst Kris, one of Freud's three most devoted apologists and promoters. The other two were Heinz Hartmann and Rudolph Loewenstein. My dream could mean that I construe HMS as a very long distance race for the true scientific base of psychiatry. I have no open competition with either Kitty or her husband, who are still legitimate and useful colleagues despite all my efforts to discredit their theories. And it is true that at several points I have given up on converting already committed psychoanalysts and focused attention instead on young students who need, as I did, to be warned about the pit of quicksand that is psychoanalysis.

What bothers me is not my relationship with Kitty, with whom I am—in fact and in fantasy—in a comfortable alliance, since, as consulting psychiatrist to the Harvard Medical Student Health Service, she often sees young women whom I have gotten to know through my William James Seminar activities.

What bothers me is the silly scenario of a marathon. I try to deal with this assumption imposed on me somehow by the dream process, but owing to the failure of my thinking capacity, I can't even calculate times or distances accurately! Naturally, an indoor track is less than a quarter mile in circumference. Even outdoors, a marathon of more than 26 miles run on a track would require more than 104 laps, a notion whose absurdity does not occur to me in the dream. I do not "wake up" to the fact that I am dreaming. Instead, I scale down the length of the race!

Reducing the length to 6 to 12 miles instead of 26 is the best my poor brain can do under the straitened circumstances of sleep. I know I cannot easily manage even 6 to 12 miles of continuous running, so I try to cut a deal with Kitty about sticking together at whatever her pace might be, sure that this will be more comfortable than mine. One minute per lap would be okay on a track that was 1/8 mile, and 1/2 minute per lap would be okay on a track that was 1/16 mile in diameter. But there is a 100 percent difference between these assumptions and those of the dream, which neither Kitty nor I seem to notice.

No wonder I am not reassured! My mind isn't working properly, and neuroscientists the world over now know why. My dorsolateral prefrontal cortex is deactivated at the same time my limbic and visual systems are turned on. Without my dorsolateral prefrontal cortex, I am as demented as any poor patient lacking a full contribution from that structure.

I turn to an activity I find easier to manage: kissing sweetly and innocently. But this doesn't work either, because it brings to the fore the importunate bevy of Hispanic women wanting to be kissed! In the past, I had a dalliance or two with Latin ladies. But none of them were Hispanic. Hispanic women were an issue at Harvard Medical School, but I was *not* involved with any of them—even administratively. Did my dream *know* that? And fear that? Was I actively avoiding Hispanic women because I might become attracted to one of them? All of these possibilities exist. And none can be disproved. Proving a negative is not something we cannot expect science to do, and neuroscience is no exception to that rule. So we will never know, for sure, why I decided to run a couple of laps and then, inexplicably, stop. The preceding formal explanation is, to me, quite sufficient and satisfactory.

I felt very lucky when I heard that our new dean was a friend of Ed Evarts. Evarts had been one of my most inspiring mentors during the two years I spent at NIMH in Bethesda, Maryland. It was there that I learned sleep laboratory techniques from Fred Snyder and became convinced that sleep and dreams could be scientifically studied to the great advantage of psychiatry. It was with Ed Evarts's encouragement that I went to Lyon, France to work on the brain mechanisms of REM with the French neurosurgeon and neurobiologist Michel Jouvet.

I knew at once that I could make a case with the new dean for his support of a career dedicated to high-risk research in this uncharted territory. After all, he was linked to Ed Evarts, my first teacher of basic (animal) sleep research. Some call these links an old-boy network, but that epithet devalues the important intellectual lineage they promote. In my case, it was gratifying to know that Dean Tosteson recognized the promise of my career, because it had been set by applying Evarts's single-cell recording technique to the mechanism of Jouvet's pontine REM generator. In other words, I wanted to know what neurones in the pons were responsible for generating REM sleep and for promoting dreaming.

When I had the Harvard Marathon dream, I had been working for eight years in the dingy back room and basement of a state mental hospital on a piece of brain tissue smaller than your pinkie! Visiting colleagues said of my lab, "Is this supposed to be Harvard?" and "Did you do *that* work *here*?" I nodded yes to both questions and then showed them how.

It is hard, even for me, to imagine how improbable this story really is. I was the apostate refugee from the psychoanalytic theory that held Harvard and most other major departments of psychiatry in its thrall. But I sought—and received—support and succor from psychoanalytically trained clinicians such as Jack Ewalt and Elvin Semrad for a mission they certainly did not understand. Most of my scientific peers thought my project would founder for technical reasons: We would not be able to record from brain stem neurones because of movement of the head, we would not be able to identify the neurones if we did succeed in recording from them, and we would not be able to analyze the data in a meaningful way. What could be more unpropitious?

But I had seen Ed Evarts's monkeys awake and moving as single-neurone recordings were being made from cells in the motor cortex. And I had seen the clock-like regularity of REM periods in Jouvet's cats, who had only a pons and medulla for a brain. To me, that meant that there was a neuronal machine of great force and reliability in that small piece of brain tissue that is called the pons.

By the time I had breakfast with Dean Tosteson in 1977, I knew my intuition had been sound. I badly needed his patient support while I rounded out the story. Tosteson was more than an Evarts crony and fan. He was also a neuropsychiatrist at heart because he had worked on effects of the mood elevator lithium on red-cell membranes. As much as Tosteson admired my neurobiological intentions, he thought I had a bee in my bonnet about Freud. He was right about the bee but wrong about the bonnet.

I had a second stroke of good luck when, in 1968, Bob McCarley fought his way into my lab over my protests that, since I didn't know what I was doing, I could ill afford a trainee coworker. Together we tried to record from brain-stem neurones and, after three difficult years, we finally succeeded. During that uncertain time, we thought a lot about what we might find. In so doing, we developed a new way of thinking about brain process and a new way of analyzing data to test hypotheses about it.

Our paradigm differed radically from Freud's. We were looking for spontaneous activity, not reflex activity. We were looking for a clock that was composed of spontaneously active neurones, not for the external energy and information that were trapped within the system. We found what we were looking for and much, much more. When we came up for air, in 1977, a paradigm shift had occurred that excited us even though we only partially understood it.

To make a very long story very short, our microelectrode probes of the cat brain stem yielded two cell groups of great interest among the many of little to no interest. The first group of neurones fired at very high rates in REM and in clusters preceding the eye movements. Many of them also fired with movement in waking. They became our leading candidate for a REM generator role. We suggested that their firing contributed to brain activation in sleep, to the shutting of input-output gates, and, above all, to the motor system activation that produced the REMs and the internal signals related to them, Jouvet's famous PGO waves. We expected to find such cells and did so relatively easily because some of them were quite large. The relationship of their discharge to REM sleep events was positive, but the two events were only loosely coupled.

Other brain stem cells had much more precise discharge coupling to the PGO waves and to the REM sleep eye movements. We suggested that they played a very specific causal role.

For two reasons, our more important discovery of the cells that turned *off* in REM sleep was unexpected. One reason was that we had found no such cells in this or other parts of the brain— nor did anyone else. Thus, the REM-off cells of the pons were unprecedented. The second reason is that the locus and probable chemical identity of the REM-off cells had led Jouvet and his followers, including us, to expect the opposite of what we found. The serotonin-containing cells of the raphe, one of our REM-off cell populations, were expected to *increase* their firing rate in sleep. And so were the norepinephrine-containing cells of the locus coeruleus; instead of directly generating REM, as Jouvet predicted, these REM-off cells seemed to us rather to permit REM to occur by ceasing to discharge.

I knew all this when I created Dreamstage. The story is in the catalogs and in the exhibit itself. And I knew all of this when I had breakfast with Dean Tosteson that spring morning at the Harvard Faculty Club next door to the Carpenter Gallery where Dreamstage was drawing unprecedented crowds. "Please let me enjoy the experimental fruits of our new paradigm," was my message to Dean Tosteson. Because he was so generous in response to this request, I agreed to work with him on his educational reform agenda. By 1984, this effort was coming to a natural end. And my dream shows it. What should we do for an encore?

The dream doesn't answer that question. Instead, it is stubbornly stuck in the unproductive milling around of preparation for the marathon. Unavailable to my dreaming brain-mind were the following salient facts:

- We were making rapid progress in our science, which took several decisive turns after 1977. One was the successful program of turning on the REM generator via the microinjection of drugs that mimic or enhance naturally released acetylcholine, the chemical we luckily guessed was running the show when released from inhibition by the REM-off cells. By giving those neurones a cholinergic chemical boost, we liberated them from aminergic bondage.

- Our discovery that the biogenic amines serotonin and norepinephrine were released selectively in waking and suppressed selectively in REM gave our model of reciprocal interaction broader relevance and enabled us to expand our conceptual horizon to waking. It made sense, intuitively, to learn that such specific and crucial wake state features as attention, learning, and memory depended on the release of the very neuromodulators we found to be absent in sleep when just these mental functions were impaired.

- The subtraction of aminergic neuromodulation, together with the enhancement of cholinergic release, suggested more searching ways of studying dreaming. It was on the basis of the neurobiological model of dreams that we constructed our new model of the mind. The Activation-Synthesis hypothesis of dreaming later evolved into the AIM model of conscious state determination.

Meanwhile, we were measuring important differences between wake and dream consciousness in several domains.

We learned that dream imagery was strongly visuomotor. That is, dreamers imagined themselves to be moving through dream space with a clarity of vision that was most intense during REM sleep. The movements were often qualitatively unusual, such as flying or skimming over a watery surface; trajectories, circular and spiral, were often experienced.

Cognitive functions were likewise distorted: Orientation to time, place, and person was deficient; reports of thinking were surprisingly rare; most significant of all was the marked deficiency in episodic memory. By an "episodic" memory deficit, I mean that autobiographical information readily available to the subject on awakening was unavailable to the subject in dreaming. In our efforts to further explore memory functions, we found that weak associations were actually enhanced in REM. This confirmed the characterization of dreaming as hyperassociative by many observers since David Hartley first emphasized this quality early in the 18th century.

When we asked subjects to score their dream reports for emotional accompaniments, we learned that elation, anxiety, and anger were prominent, while socially learned feelings such as shame and guilt were rare.

I will explore each of these findings in discussing other dreams. I mention them here to establish the nature and power of the bottom-up approach to distinguish it clearly from psychoanalytic and personal meaning approaches to thinking about dreaming. We use this approach to study dreaming scientifically and emphasize the constraints it places on the interpretative endeavor.

Because so many of the cognitive deficits of dreaming relate to memory, we naturally wondered what the mechanism of dream amnesia might be. Three possibilities derived from classical studies of memory suggested themselves. The first was our own discovery that serotonin, a molecule shown to be essential to learning in Eric Kandel's studies of learning in snails, was unavailable to the human brain in REM. The second was the finding that the dorsolateral prefrontal cortex, shown by Patricia Goldman-Rakiç to be crucial for working memory in monkeys, was not activated in REM. And third, we wondered if the flow of information from hippocampus to cortex was altered. The hippocampus has been known to be essential to autobiographical memory since its surgical removal leads to complete amnesia. Recent neurophysiological work by Gyorgi Buszaki has suggested that, in REM, hippocampal information becomes unavailable to the cortex, which, as I have just explained, is deprived of a crucial chemical, serotonin, and regionally inactivated in REM sleep.

Our research team, led by Roar Fosse, struggled for a long time to find a way to show the obvious impairment of memory within dreams. Although this formal feature is robust, it is not widely recognized. We forget almost all our dreams. Aided by my directed search and journal recording technique, I have accumulated about 350 reports—some quite detailed, but others, like this one, admittedly incomplete. At age 70, assuming I have 5 REM periods per night, I must have had at least 127,540 dreams. My documented recall is less than two-tenths of 1 percent.

Amnesia for dreaming is so pronounced that it seems to have obscured the memory defect within dreams. Knowledge of Norman Geschwind's recent death is unavailable to my dreaming brain-mind. In some dreams, characters appear alive who are in fact dead, and the dreamer doesn't notice. We think these defects in memory, orientation, and critical thought are linked, just as they are in delirium, and for the same reason: A change in neuromod-ulatory balance causes both. One is spontaneous, natural, and reversible. The other is drug-induced, artificial, and remediable.

In our dream memory experiments, we asked our subjects to record these dream recollections and then perform their own search for memory sources. As in my Harvard Marathon dream, it was easy to find such sources, but critical evaluation of these sources revealed how very incomplete their representation was in the dream. Our subjects were instructed to compare features of their dream memory source with those of their waking recollec-tion. Of the more than 300 reports we collected, we found only three that reproduced in dreams the time, place, person, and action features easily available to the awake mind. We concluded that brain memory systems in REM sleep work in a very different way from that of waking.

Why didn't we predict what brain imaging would later show? When we first formulated our activation-synthesis model of dreaming, we did propose that the visuomotor imagery of dream-ing depended on selective autoactivation of the pons, lateral geniculate body, and occipital cortex. And we did predict that dream emotion might result from autoactivation of limbic fore-brain structures that we already suspected might mediate emo-tion, especially anxiety, a dream emotion that was prominent in our early work. But we missed the boat entirely with respect to the cognitive difficulties of the dreaming mind.

It was not until the mid-1990s that scientists succeeded in obtaining good PET images of the human brain in sleep for com-parison with waking. PET images refer to brain structures "light-ing up" to convey via a color code the intensity of the regional blood flow differences that are measured when a positron-emitting tracer is injected during REM. Via a subtraction proce-dure, the color code can show the blood flow intensity in relation to some other state, usually waking. Because both waking and REM are states with high levels of general activation, it is difficult to discriminate them using brain wave patterns alone. The EEG is too crude and insensitive. But PET scans show these differences clearly.

In addition, the brain areas that lit up, many of which we pre-dicted, were brain areas that were underactivated in REM. It was this finding that we did not anticipate. We felt that the aminergic demodulation was enough to explain the cognitive deficits of dreams—such as my inability to calculate the distance of the dream marathon or to exercise any self-reflective awareness of the absurdity of the dream's actions.

We never guessed that REM sleep entailed a specific inactivation of the cerebral cortical seat of executive cognitive functions. By executive functions, I mean such critical brain-mind management tasks as working memory, directed thought, and self-reflective awareness. These are precisely the functions that dream phenomenology was trying to tell us to attend to. The dorsolateral prefrontal cortex has been known to be the seat of executive functions for some time, so we cannot plead ignorance to explain our failure to predict this finding.

As embarrassed as we were to have missed this prediction, we were delighted to learn that it had been picked up by the brain imaging studies. Why? Because it helped establish an objective basis for the cognitive deficits that our psychological probes of dream memory, orientation, and thinking were documenting. In any bidirectional mapping effort between the brain and our consciousness of its activity, we need to pursue each approach with an eye and an ear to the other.

Fortunately, many other scientists were working on the changes in memory function during sleep. At sleep onset, we often experience a replay of unusual daytime activity such as skiing, boating, or, in Robert Frost's case, apple picking. At the University of Arizona, Jonathan Winson and Bruce McNaughton were able to show that hippocampal neurones involved in wakestate learning were reactivated in sleep. This approach was taken even further by Matt Wilson at MIT, who showed that rats that learned a maze when awake relearned it during sleep.

My colleagues Bob Stickgold and Matt Walker have also demonstrated that the learning of a procedural task is protected by sleep and can even be enhanced by it. The relationship of all these findings to the conscious experience of dreaming is unclear, but one plausible suggestion is that the formation of new memories—memory of the dream itself, for example—is sacrificed in favor of a more practical and serviceable embedding of previously learned material.

At the time I had my Harvard Marathon dream, I didn't even know that imaging of the human brain in sleep would be possible.

If I had, I would have dropped everything and run to the scanner. Or would I? We were having so much fun turning REM on and off with drugs that I might have stayed home and minded my little store. It was certainly profitable to do so.

Brain imaging brought to the fore another fine point that had escaped us. While we were advocating a pons-to-forebrain direction for state-determining processes such as neuromodulation and PGO waves, we were in no position to say which came first in dream plot construction: the chicken of the dream imagery or the egg of dream emotion. We are now in a strong position to suggest that it is the egg that comes first. That is, we feel a certain way in dreams, and then we see imagery that is in keeping with that feeling. We know that this kind of process can also accompany strong emotion in waking. If we are anxious, or sad, or in love, we tend to see and think apprehensively, or melancholically, or amorously. In REM sleep, we also know that the emotional brain is selectively activated.

We had our first clue that dream emotion drove dream cognition in our early studies of dream plot bizarreness. Dream emotion was always consistent with dream plot features, no matter how out of kilter those plot features were with each other. In scene 1 of my Marathon dream, the faces of Deans Tosteson and Federman map well onto my feeling of anxious apprehension about them even though they are presented in an absurd manner and change back and forth from one to the other. Similarly, my sense of comfort and intimacy with Kitty Beck holds scene 2 together despite the centrifugal forces of this unlikely event and my inability to reason about it. Anxiety (scene 1) and affection (scene 2) give way to playful seductiveness (scene 3) when I am tempted to kiss the Hispanic women.

When we look at the pattern of brain activation revealed by PET in REM, we are struck by the proximity of the limbic structures to the pons and hence are led to wonder if it is not subcortical ponto-limbic activation that determines dream emotion. We also wonder whether the visuomotor images and plot features are fit to them as best the sleeping brain can, given its isolation from the outside world (its anchor in waking) and its lack of cortical control (its compass in waking). This hypothesis resonates with what I call the emotional salience of dreams. Freud would call it primary process and ascribe its presence to a relaxation of vigil over the id by the ego. We agree on the nature of the process up to this point. But instead of disguise and censorship, the direct revelation of feelings is impressive to us. Disguise and censorship are weak if they are present at all.

All the disparate elements of my Harvard Medical School dream are connected to the Class Day scenario, and they clearly reflect my confusion and ambivalence about the interaction of my impulsive, artistic, and hedonistic impulses with the politics and decorum of institutional life. In my dream, important people are present with whom my complex relationships are played out.

That there are symbols is indisputable. The marathon is a metaphor for Harvard Medical School. I run, but I never quite know why, with whom, or for how long. The paper bag with the deans' faces printed on it represents the promotional aspect of our professional lives. But these symbols do not disguise meaning as much as they reveal it. As in the Dreamstage dream, the unification of disparate and bizarre images is achieved by linking them via emotional salience rules.

One final point about the spectrum of dream emotion: From a Darwinian perspective, the emotions of anxiety (which engenders wariness), elation (which engenders affiliativeness), and anger (which engenders defense or attack behavior) are highly adaptive. Be careful; find a mate; and drive off competitors and predators. These emotions are the building blocks of survival and procreation that Darwin has taught us are the twin goals of life. An attractive theory, first put forth by Jouvet, is that REM sleep provides an opportunity to rehearse the survival emotions and behaviors. The Finnish scientist Antii Revonsuo recently echoed this idea. To this, we can add the hypothesis that REM sleep emotions also provide an organization structure for memory and account for the striking emotional salience of dreams.

Chapter 3

Lobster Brain
The Visual Brain as Image Synthesizer

I am in a dream hotel with an uncertain room assignment and nondescript Samsonite luggage. The presence of my scientific colleagues Ralph Lydic and Barry Peterson indicates that we are probably at a meeting, but this is unclear, and it can't be our luggage (it is too industrial) or our room (it is too small, and there is too much luggage and no furniture). I decide to leave, but luggage blocks the door. Frustration ensues.

The scene changes to a smorgasbord-like banquet table, which is cluttered with platters of elaborately prepared food. I am attracted by the lobster because it is huge (almost half human size) and a deep, bright red.

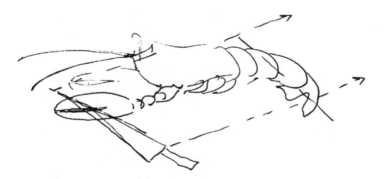

A hostess (who is not really seen) is serving and offers to cut me a piece of lobster. She begins by slicing through the left claw (which offers no resistance to her knife). Her cut continues through space in a direct plane until it encounters and slices the tail. The two pieces are joined—as in a drawing of conic sections—and I am served both with one miraculous gesture.

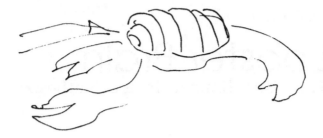

The hostess then asks if I would care for some lobster brain. I think to myself that lobster brains (even in larger-than-life lobsters) are too small to make a meal. But when I look back at the lobster, I notice that the brain (which has moved to the center of the back) is huge and elaborately prepared in wedge-shaped slices. I am told I should use chopsticks (at which I am adept).

I select a piece and am amazed to note that it has a cream-colored border and a deep red center. On the cut surface of each slice is a bas-relief of a plant or tree-like form. "What wonderful things they can do with food," I exclaim with my characteristic gastronomic enthusiasm. Then I become disturbed because I notice that the lobster has no aroma and, by implication, no taste. Since this is a dream feature of some interest to me, I decide to wake up and take note of it.

At the time of this dream, I had become interested in dream bizarreness and was struggling to find a way to characterize it. I therefore was motivated to awaken and to take careful note of my experiences. In this case, I was especially interested in what appeared to be the almost-complete exclusion of taste and smell from dream reports.

When I noticed that the elaborate dream lobster dish that was being served had no aroma, I woke up! Was I aware I was dreaming? Was I lucid? Probably not, because I very likely would have noted that fact in my report. But if I was *not* lucid, how could I decide to wake myself up? Obviously, there was a sufficient degree of priming to tip off my dream consciousness that an aroma-free lobster dish was of scientific interest, and it was enough, if you believe my report, to prompt a voluntary self-awakening.

This set of issues raises important questions about self-recording and self-analysis as well as presenting research opportunities that have yet to be exploited by neuroscientists. For example, how can I be sure that this dream was spontaneous, as opposed to being induced by presleep autosuggestion? I cannot. But since I don't remember giving myself a presleep prompt, it seems more likely that I was primed to notice this feature rather than to dream this particular dream. Another reason for assuming that this hypothesis is correct is that a lack of insight permeates most of the report, especially when I accepted it as real despite a host of improbable features such as the enormous soft-shelled lobster and its larger-than-life brain.

As an example of the scientific opportunities offered by our capacity to program our dreams, consider the prospect of my becoming lucid in this dream and, rather than waking up, of actively pursuing the taste-smell issue. Why don't I take a dream whiff of the slices, or sample the whole dish, or the sauce to see if it was indeed odorless? Or I could take a bite and see if my dreaming brain could taste the lobster.

Positive results with either of these experiments wouldn't really prove anything, because in order to achieve dream lucidity, the brain must assume some of the features of waking. Those features could well include taste and smell, modalities that are not normally present in non-lucid dreaming. A negative result would be more interesting. If I could not smell or taste the lobster, even when I indulged in active dream behaviors such as sniffing and swallowing, I would be obliged to accuse myself of self-deception to explain away the result.

In the preceding chapter, I explained that the frontal cortex of the human brain, which is known to support executive cognitive functions, is selectively inactivated in REM sleep. This physiological fact may well explain the central psychological deficits of dream consciousness—our inability to reason, to direct our thoughts, and to have insight about our state. We all think we are awake when we dream. And this is a mistake. Can it be corrected?

The short answer is yes. But the longer answer is so interesting that the full story must be told. We have all experienced glimmers of doubt about the veracity of our dreams. "This is too crazy to be true," we say. Then we are pulled back into the delusional sea of false belief by a new rush of dream imagery. It is the spark of this fleeting disbelief that we fan into the fire of lucidity when we become conscious and aware that we are dreaming even as the dream proceeds.

In the example provided by my Lobster dream, I was not fully lucid, but I did wake up. Waking up is quite common when lucidity is attained, suggesting that lucid dreaming might occur in a hybrid state, with some features of waking and some features of dreaming. For someone who wants to cultivate this fascinating dissociation of dreaming, it is possible to increase its probability of occurrence by priming the brain-mind to notice dream bizarreness via presleep autosuggestion. By placing a journal or tape recorder on your bedside table, you can perform this experiment on yourself.

If you are young—say, under forty (or, better yet, thirty)—you are very likely to have success. As with all of life's pleasures, older people have to try harder. In my distant youth, sleeping at odd hours, I could become aware that I was dreaming and change the dream plot at will. Or I could will myself awake, the better to remember my lucid dream. Then I could go directly back to that dream or another of my own devising. Sad to say, this talent has now gone, along with so many others.

Why should athleticism of the mind be any different from other physical sports? Wisdom should increase with age, so the ease of inducing altered states of consciousness should therefore increase, not decrease. As many have seen in dear but crotchety grandparents, this is not always the case. So, make hay while the sun shines, and consider, with me, other experimental possibilities.

How does the brain support two opposing states at once? How can the brain be both awake and asleep, dreaming? It cannot if we consider the brain as a whole. But if, instead, we consider the brain to be an ensemble of regional parts, as the PET experiments show it to be, we have less difficulty imagining what is going on. One part of the brain wakes up while the rest remains asleep. Why not? We know that whales and dolphins do it, so why can't we?

The demonstration that dolphins sleep on one side of their brain while the other remains awake was made by Lev Mukhametov, a Russian neuroscientist who had earlier worked on the cellular physiology of sleep in Pisa, Italy. At the Crimean Marine Biological Station on the Black Sea, Mukhametov was able to record the EEG of dolphins in a restraining tank. When the left brain emitted high-voltage slow waves typical of deep sleep, the right brain might emit the low-voltage fast pattern of arousal, and vice versa.

The expression "asleep with one eye open" says it all. On the falling-asleep side, it is possible to be in two states at once. So why not on the waking-up side? Suppose you were in the scanner while your mind was in two states at once. What would we expect to see? Signs of waking in some brain regions and signs of sleep in others. In the case of dream lucidity, I would predict reactivation of the selectively inactivated frontal cortex, which then gets to watch, attend to, and finally influence what the brain stem and visuomotor forebrain (still in REM state) are doing! We would also not be surprised to see diminished activation of the limbic system and to see that dream emotion subsided during lucid dreaming. I say I would not be surprised because the frontal lobes and the limbic brain are in a constant battle for control of the mind. Here the seats of reason (Freudians read ego and superego) struggle to regain dominion over the seats of emotion (Freudians read id). In this case we have little difficulty updating Freud's important sensitivity to the reality and power of primordial feeling in competition with reason and social demand.

No wonder intermediate states of the brain-mind such as lucid dreaming are so rare. And so fragile. It would not make survival sense to have the brain be freely dissociative. The fact that the brain-mind is normally committed to one and only one state at a time is a great advantage. The social and economic handicap of pathological dissociation clearly shows this to be true. People whose waking consciousness is subject to intrusive dream-like perceptions cannot function competitively. Darwin's motto, "the survival of the fittest," places the mentally ill at grave risk. Charles Darwin was a great psychologist as well as a great biologist. His principle of serviceable habits has been picked up by modern scientists who focus on the competitive advantages of being conscious (Gerald Edelman, Antonio Damasio, and Giulio Tononi) or being linguistically communicative (Stephen Pinker).

Scene 1 of my Lobster dream is a classic dream scenario. Although I am a frequent traveler, I am sure that I have been in many more dream hotels than real-life ones. Why hotels are so often chosen as settings for my dreams I don't know, but I speculate that it could have something to do with the anxiety that sleeping in an unfamiliar place naturally inspires. Hotels are also attractive dream plot settings because anxiety is a common dream emotion; dream anxiety teams up with the confusion about spatial location that is practically universal in dreams. In other words, dream bizarreness is formally compatible with dream hotel settings.

Studies that have been under way in our laboratory since 1985 have shown that dream bizarreness can be reduced to orientational instability. The details of my Lobster dream that support this hypothesis: I am not sure of my room assignment; I see luggage but do not recognize it as mine (in fact, I know it is not mine!); my colleagues, Ralph Lydic and Barry Peterson (who are real people and appear as such in the dream), suggest that I am at a scientific meeting, but in spite of their presence, that fact cannot be established for certain. I feel confused by the lack of orientational data and decide to leave but cannot because luggage blocks the door. The frustration that ensues is of almost nightmarish intensity, but I don't wake up. Instead, there is a radical scene shift.

How we discovered the disorientational essence of dream bizarreness is an instructive tale. For years—centuries, even—

people have awakened from sleep and said to bed partners, family members, friends, and high priests, "I had the strangest dream!" And professional psychologists have quantified this strangeness or bizarreness as parts of their scales of "dreaminess." But no one, before us, tried to unpack and analyze this robust formal property of dreams. How could we define and measure dream bizarreness in order to understand it in terms of altered brain function?

Freud theorized that the surface nonsense of dreams was the result of sleep-protecting work by the ego to censor and disguise the unacceptable wishes that instigated dreaming when the ego let go of the id. As unlikely as this idea seemed to me, we had no alternative except for the 19th century neurological nonsense notions that Freud had ridiculed as vague (which was true) and unhelpful (which was false). But no one, including Freud, his predecessors, and followers, had really defined dream bizarreness psychologically. What was it? Was it robust? And was it analyzable?

Our answer to all these questions is a resounding "yes." Many psychologists have been taught to think of dream reports as consisting of only "manifest" content. The manifest content was supposed to hide a deeper or "latent" content that was the real truth of the matter. This paradigm helped retard a descriptive approach to dream bizarreness. It is embarrassing to admit how long it took our own lab to see that dream bizarreness always consisted of either plot incongruity (things that didn't belong together were stuck together) or plot discontinuity (dream times, places, persons, and actions changed without notice). Dream discontinuity can be radical, as when scenes change abruptly, and such bizarreness is easily detected because it is so global and so gross. Most people now agree about this. Scene changes do occur, but they are relatively rare.

When we focused on more fine-grained incongruity and discontinuity, we found it in practically every line of REM sleep dream reports. It was fine-grained dream bizarreness that we wanted to measure and explain. Because it is subtle, it was more difficult to detect, but trained observers could do so with high levels of agreement. We called this fine-grained bizarreness microscopic because it was so easily overlooked. We concluded that the continuity of time, place, and person were violated by the dreaming brain and argued that this microscopic orientational instability was organically determined, just as it is in clinical delirium. The disorientation of dreams is normal, not pathological. This was a bitter pill for some of our colleagues to swallow. They kept insisting that dreams were banal and were as well organized as waking mental content.

The scene switch in my Lobster dream from hotel room to function room makes some sense because one of the first events at scientific meetings is the opening reception. But our studies of dream splicing, where reports like this are cut apart at just such scene changes, show that blind observers would *not* be likely to identify the lobster dish of scene 2 as a sequel to the hotel room of scene 1 because the shift in subject matter, emotion, and tone is *so* radical. The scientific meeting context that appears to link these two scenes is absent from the second scene. It is only when we *know* that scene 2 follows scene 1 that we can come up with explanations that are almost certainly projections rather than correct analyses of brain-mind causality.

So why do dream scenes change? Brain physiology changes rapidly in sleep, and it changes more radically in REM sleep. Eye movement clusters come and go, and with them autonomic storms build and subside. One startle response after another is inferred from the clusters of PGO waves seen in animal studies. We call these moment-to-moment fluctuations phasic activation events. Our home-based Nightcap recording studies suggest that each REM cluster can generate its own dream scenario. Thus, there may be at least five cognitively and emotionally separate dreams in any given REM period. For all we know, we could be having twenty-five to fifty separate dreams each night, and we would never know it, because awakenings are so rare.

When the dream is going on, we are often surprised, or even startled, by the turn of events. I am made anxious when I see the peculiar luggage, the too-small hotel room, and my colleagues, Lydic and Peterson, standing together. My surprise is too weak to register as such, but it is there at a low level most of the time that I dream. Occasionally, it becomes so prominent that I notice and record it. Like when the scene changes from the hotel room to the dining room where the lobster is being served.

Neurophysiology tells us that the forebrain is radically deactivated in deep NREM sleep. NREM sleep is characterized by unconsciousness and high-voltage slow EEG waves. It occurs most prominently in the first half of the night, as if it were a recovery function related to time spent awake. In the PET scan, as well as in our consciousness, NREM is "lights out in command central." Overnight, the brain is progressively reactivated. As successive NREM epochs become lighter and successive REM epochs longer and stronger, the probability of dreaming being associated with them increases. Because these events occur over time periods of many minutes and hours, brain scientists call them tonic events.

But brief, episodic "phasic" events occur in NREM sleep, too. These include spindle waves and K-complexes in the EEG. Their psychological concomitants have not been well studied, but it has been speculated that they have to do with learning. In REM sleep, such phasic events include the eye movements themselves, and their hypothesized signal associates the PGO waves so easily recorded in animals. Clearly, phasic REM is physiologically more activating to the supposed seat of conscious experience in the forebrain than tonic REM.

I speculate that REM sleep dreaming is constantly novel and unpredictable because the brain is running its startle program in an uninhibited way. What might this mean? It might mean that a good way of getting attention, a good way of provoking emotion, and a good way of ensuring learning is by means of surprise. I remember exactly what I was doing when I heard that President Kennedy had been shot. My emotional brain was powerfully activated when I heard the news. It is true that I don't remember most of my dreams, but that doesn't mean my brain is not changed by REM sleep. Dreams are full of surprise, and REM sleep is full of startling events.

Consider scene 2, the lobster banquet. Bizarreness still holds sway even though the plot has changed beyond recognition. The lobster attracts my attention because it is huge, "almost half human size," and a very deep red. I have seen some big (2 1/2 pound) and even giant (5 pound) lobsters in my time, but never one that was anywhere near half human size! That would have to be at least 50 pounds, 10 times my largest real-life lobster. And the exoskeletons of boiled lobsters *are* red. But their color is usually orange-red, not a deep red like this one. In sleep, and especially REM sleep, my autocreative brain is not limited by these restraints. In dreaming, it would seem that anything goes.

It is also typical for dream characters (like the hostess who is serving me lobster) to be unidentified or even, as in this case, unseen! The dream plot needs someone, anyone, to fill this function. That someone is invisible and performs feats that are even more magically outlandish than the lobster itself! I have never seen a lobster cut in this way, and I never expect to, except in my dreams. Even very sharp knives don't easily cut lobster claws, and lobster tail carapaces are even more obdurate. Why would anyone want to slice a lobster in this manner, anyway? That I am enjoying the hostess's tour de force is clear from my characterization of the serving of the two joined pieces of lobster as a "miraculous gesture." This is a dream of high comedy, pleasure, and elation. I marvel at it even now.

The hostess next asks me if I would care for some lobster brain. This question activates my reasoning power a bit, but not enough to give me the insight that I must be dreaming. And not long enough to prevent me from creating an even more preposterous illusion. I figure that since even large lobsters have very small brains, her offer is absurd. But like the larger-than-life lobster, the dream lobster brain suddenly becomes huge, as if to suit the needs of the invisible hostess's offer and my ravenous appetite. The lobster brain has also moved to the center of my visual field, where it is seen to have been prepared in wedge-shaped slices.

The plastic possibilities of the dreaming brain are apparently endless, as this sequence reveals. "Is this the way great chefs get their ideas?" I wonder as I contemplate how the dream lobster brain wedges have been prepared. Such an elaborate design could, I suppose, be imposed on medallions of lobster tail, but I have never seen anything like this, even in the fanciest restaurants of the world. My pleasure reaches its peak, as I exclaim, "What wonderful things they can do with food!" I am still in the grip of my dream hallucination. This grip breaks only when I realize that such a handsomely prepared lobster dish should have an aroma, but it doesn't! My unusual noting of this discrepancy leads to my waking up.

Recognizing the autocreativity of REM sleep dreams is important to our view of ourselves as humans. Although we are automata in the sense that to survive we must eat, we must sleep, and we must maintain body temperature, it is delightful to learn that in managing these functions without our thinking about them, our brains are capable of the most ingenious inventions. Some of these are throw-away items.

In the words of the philosopher Owen Flanagan, dreams are the spandrels of the mind. They are mere decorations to delight us rather than the weight-bearing columns or space-spanning arches that are the real stuff of architecture. In this analogy, the dream is the spandrel and the REM sleep restoration of thermal equilibrium is the column. Again, we see how the brain-mind has evolved in this case to permit both basic housekeeping functions and decorative diversions. I think the late Stephen Jay Gould, who originated the spandrel idea, would have been pleased by this point.

Dreaming is simply delightful (or horrific) in the way that all works of art are delightful (or horrific). We are drawn to them because they highlight reality in an emotionally compelling way. I have heard it said that "To tell a dream is to lose a reader," but it has never seemed that way to me. I love to listen to other people's accounts of dreams, and I love to share my own with them. No one can resist interpreting them, just as we cannot resist judging any cultural artifact. But, as scientists, we stop short of presenting a rebus for reductionist analysis of dreams.

When we change our view of dream interpretation from a many-to-one to a one-to-many paradigm, we recognize that artistic processes are natural to all of us. Life is short. And art is long. Society and civilizations are built on the creativity that goes with the flow of imagination that comes to the surface in our dreams.

Scene 1 never develops a story because no orientation frame can be successfully established. The only building blocks are my colleagues Lydic and Peterson. Both of them are physiologists who have contributed important work to sleep science, and they are closely enough associated to one another to explain their copresence in a scientific meeting dream. Even in waking, I could invent dream scenarios to accommodate them. This is true even though I have no recollection of ever having been with them together in a hotel room, at a scientific meeting, in my lab, or any-where else. So why are they together in my dream?

Barry Peterson worked with Victor Wilson at Rockefeller University on reticulospinal neurones in relation to vestibular function. I met him through Terry Pivik, who invited Barry to present his work at a sleep research meeting in Jackson Hole, Wyoming in the late '60s, long before I met Ralph Lydic. Peterson and I hit it off, and I kept in touch with him for many years after he moved to Northwestern University in Chicago, where he worked on sensorimotor integration. It was clear to me from the start that Peterson's research paradigms and techniques were very useful to us sleep scientists, and he generously encouraged my work with Peter Wyzinski on reticulospinal neurones. These large brain stem neurones could be identified by antidromic invasion and could be shown to be activated in REM. In transition from NREM sleep to REM sleep, the executive motor system of the upper brain is turned on, and the motor output channel of the spinal cord is turned off. It was while doing one of these experi-ments that Wyzinski made the accidental discovery of cells in the locus coeruleus that turned off in REM.

Barry Peterson was an important colleague in many ways. He helped me see that the brain stem circuits that interested both of us had properties of great interest. Through Barry, I reread, and later met, the great neuroscientist Rafael Lorente de Nò, who wrote his classic paper on the vestibulocular reflex in 1933. The fact that eye position, head position, and trunk position (or pos-ture) were all integrated in the pontine brain stem and cerebellum meant that the activation of those circuits during sleep could con-tribute to our elaborate and continuous sense of movement in dreams. The same theory could also explain our sometimes impossible dream movements. The recognition that the sleep acti-vation of these visuomotor circuits occurred offline (when the brain stem was disconnected from the external world) helped me grasp their possible significance for dream formation.

Identification of at least some of the neurones we recorded helped us gain the respect of many mainstream neuroscientists who still remained within the safety of the Sherringtonian reflex paradigm. I would never have asked Wyzinski to come from

Mircea Steriade's lab in Quebec had I not wanted to show that we too could identify reticulospinal neurones using the classic technique of antidromic invasion. This approach allows long-axoned neurones to be backfired at some distance from the recording electrode and establishes, by ironclad criteria, that the neurone being recorded projects to the stimulation site.

But it was Wyzinski's inadvertent discovery of REM-off cells in what our histology showed to be the locus coeruleus that led us to the scientific gold. Based on the inadvertent discovery of the REM-off cells, we were able to create a plausible model for explaining the endogenous changes in excitability of reticulospinal neurones and other cells in Lorente's vestibulocular reflex circuit. To make this point clear, ask yourself this paradoxical question:

When is a reflex not a reflex?

"Never," say the Sherringtonians, and the Freudians echo the cry.

"In REM," say the state control neurobiologists who know that reflexes can be annulled, enhanced, or even reversed when the brain changes state.

As soon as I heard the regular tap, tap, tap of Wyzinski's outlaw cells, I knew we had found a new breed of neurone in the brain, a neurone that identified itself via its slow, regular discharge pattern during waking hours. We followed these outlaw cells through several sleep cycles to be sure that our electrode hadn't moved out of the cell's electrical field.

At the time we made this discovery, I didn't know enough neurobiology to realize that these slow, metronome-like firing patterns were typical of cells that were both pacemakers (that is, self-excitatory) and purveyors of neuromodulators (such as serotonin and norepinephrine). I did think, from Jouvet's work, that locus coeruleus neurones were not supposed to behave as they did if they were REM generators. They should have fired more in REM, not less. It was not until I saw the histological location of these cell recordings that I knew Jouvet and all the rest of us had the rule backwards. The locus coeruleus was permissive, not executive, of REM. It needed to be inactivated, not activated, for REM to occur. "Eureka" is a tame word for what I felt at that moment.

Ralph Lydic, like several other very helpful junior colleagues, fought his way into my lab shortly after we had made the locus coeruleus REM-off discovery. Bob McCarley had developed the reciprocal interaction model of sleep control, and we had thrown down the dream gauntlet to the Freudians with the activation-synthesis hypothesis of dreaming. So Lydic fits, in a way, into my dream hotel room along with Barry Peterson.

Ralph, like Barry, was a "real" neurophysiologist, not a clinician-turned-neurophysiologist like McCarley and me. But Ralph had an unusual sensitivity to and appreciation for our interdisciplinary efforts. Lydic was trained at Texas Tech University by John Orem, who was pioneering the study of sleep-dependent changes in respiratory neurones. Before coming to my lab, Ralph worked with Martin Moore-Ede at Harvard on the neuroanatomy of the circadian clock in the hypothalamus.

When Lydic joined our group, we had just confirmed Dennis McGinty's findings that there were REM-off cells in the dorsal raphé nucleus, too! Since these neurones are serotonergic, this finding fit neatly into the reciprocal interaction model. The brain self-activates in REM sleep when enough time has passed for the noradrenergic neurones of the locus coeruleus and the serotonergic neurones of the raphé to be completely silenced (as we now know, by active inhibition). This shutdown allows the cholinergic and glutaminergic neurones of the brain stem to fire up and to trigger REM by driving the forebrain activation, eye movements, and muscle tone inhibition that block motor output.

We wanted to test the mathematical aspects of the model, and we had good activity curves from our many REM-on cells, but nowhere near enough REM-off cells. It fell to Ralph to design a system for the long-term recording of the raphé neurones. Working with great skill and self-discipline, he was enormously successful. Lydic's raphé neurone data strengthened our hand considerably and allowed the model to gain credibility on the theoretical side. But remember, reciprocal interaction was also a synaptic model with numerous testable predictions using neuropharmacological tools.

The discovery of REM-off cells in the locus coeruleus and raphé nuclei in 1975 led to the reciprocal interaction model and to all that flowed from it thereafter. In the light of the reciprocal interaction model, Lydic conducted our most successful studies of the serotonin neurones. Like Barry, Ralph was a classically trained physiologist. So it *does* make sense to find myself at a scientific meeting with these two guys, but only in a dream, when time is ignored in favor of other important aspects of association!

When I had this dream in 1985, I was coming under strong attack from my scientific peers over the role of reticulospinal neurones in REM sleep generation. Many of them thought we had overemphasized the role of the pontine giant cells in REM sleep generation. Some even claimed that REM sleep activation was entirely nonspecific. For this reason, Lydic and his wife-to-be, the behavioral pharmacologist Helen Baghdoyan, and I wrote a monograph for open peer commentary in the Behavioral and Brain Sciences. It was published in 1986. This paper helped put our program in perspective and regain funding from the NIH.

At the same time, my personal life was in some disarray. Some of the reasons for this discovery may be gleaned from a closer look at the content of scene 2. To my eye, three items warrant comment from the point of view of emotionally salient memory. One is the lobster itself, the second is the lobster's brain, and the third is the gastronomic-culinary aspect.

Lobsters have always fascinated and attracted me. They are exotic, beautiful, and tasty creatures. I remember as a child of 8 or 10 being offered the small legs when my neighbors, the Constantine Allens, ate boiled lobster at picnics in their backyard across the street from our house. Constantine Allen was a Greek who had changed his name. He had five daughters. One of them, Connie, played doctor with me and later became a successful nurse. My parents said we couldn't afford lobster, which naturally increased my longing for it. As I grew up I found ways to get lobster and to eat the whole thing, not just the little legs.

My first real lobster orgy was in Maine with my psychologist mentor, Page Sharp, who bought them wholesale when he ordered food for his 180-person summer camp. Sharp, better known as Cupie because of his amorous nature, taught me to hide lobsters for our clandestine cookouts across the lake. So lobsters, without being in any sense sexual symbols, are linked in my mind to sensuality.

I never ate lobster brain. But I did eat many other kinds of brain—especially calf brains or *cervelles au beurre noir* that were prepared in the small café near Jouvet's lab in Lyon, France. After an all-day procedure or when I was working late at night, I would go to the café and order a dish made of the very object of our study. As I ate the brain, I could review the anatomy of its still easily recognizable parts. And I could do this without either bravado or squeamishness. So it makes sense to eat lobster brain in my dream, especially since I could both enlarge and embellish it so freely.

Eating lobster brain fits into the category of my exotic yearning for all things Gallic. As for the striking method of cutting and serving the lobster, my interest in the precise representation of complex structure is connoted in my dream reference to "conic sections." As a student at Loomis School (1949–1951), I took courses in mechanical drawing and excelled in them. I especially enjoyed the most difficult three-dimensional projects—including conic sections—that came at the end of the course. The combination of visualization and geometric mathematization came in very handy when Bob McCarley and I were exploring the brain with microelectrodes and had to make maps that converted Cartesian coordinates into polar coordinates.

Many of our neurobiologist colleagues, including Allan Selverstone and Ed Kravitz, worked on lobster brains. The joke then was that they would fish out the tiny brain, spend the day studying its neuronal activity, and then eat the rest of the lobster for supper! I have no idea if this story is true or apocryphal, but it always amused me. Another association that might be far-fetched is that many invertebrate physiologists were interested in the giant motor neurones of the brain stem that commenced escape flexions of the lobster's tail. These evolutionary precursors of our own reticular giant cells could help explain the Barry Peterson-to-lobster brain transition.

Ever since I met Page Sharp in 1948—and especially after I went to England and France in 1951—I was passionate about good food and its exotic preparation. In another life, I could have been a chef. In my dream life, I come as close as possible to that ambition. Just before this dream occurred I had eaten an exotic dinner of raw lobster (sashimi) prepared by my Japanese artist friend Kaji

Asò at his tea house dining room on Symphony Road (hence the chopsticks!). Kaji often challenges us by preparing even more exotic body parts than lobster brains. His raw sliced calf testicles are tough to chew despite extensive marinating.

Less provocative, but equally artistic in the matter of food preparation and presentation, was our French housekeeper and mother surrogate, Chantal Rode, who lived with us for 23 years (1972–1995). She is the only person I know who could have made the food decorations described in the dream! Chantal came to us via our Bordeaux friends Jean Didier and Yveline Vincent, who are mentioned in the Dreamstage dream discussion. After living with us, Chantal became a very successful restaurant hostess, first at the Museum of Fine Arts and then at the downtown Harvard Club.

It is one thing to show how a dream report can trigger the recollection of diverse memories that demonstrate parts of the vast associative networks accessible by the activated brain. It is quite another to assert that these networks actually contributed information to the construction of the dream. In other words, the utility of dream discussion in psychotherapy may gain more from the fact that some dream elements are connected to many aspects of our experience rather than the reverse. In other words, the connection is one-to-many rather than many-to-one. Many-to-one is the Freudian way. It is radically reductionistic in holding that "the real meaning of this dream is…". The many-to-one interpretative stance forces us to accept a very limited and stereotyped set of psychodynamic principles. The one-to-many paradigm is, by contrast, open and expansionistic, although no less deterministic. It says that dream contents are hyperassociative but not overdetermined, as the Freudians insist. In my view, the Lobster dream is a many-splendored image, not a symbol to be decoded reductionistically.

What I have tried to show here is only that even very small dream fragments (scene 1) or very exotic, bizarre, and detailed dreams (scene 2) can be used as starting points for psychologically sensitive autobiographical reconstruction. This can be done without any assumptions about disguise, censorship, or symbolism of the Freudian kind. In my Lobster dream, the lobster is a lobster. The dream lobster leads me in many interesting directions. If I were to hazard an interpretation, I would say that the Lobster dream pulls together many emotionally salient parts of my life. I love exotic animals, I am fascinated by their brains, and I love to eat! The dream shows how all these interests are linked deep in my brain's memory system.

There is a deeper connection between the dissociation of hybrid states and the hyperassociation of REM. In order to achieve either or both, there must be some loosening of the usual constraints on the cognitive networks in the brain, and sleep affords just that. We now know at least one good reason why this happens: The aminergic modulatory system, which provides a generous supply of chemicals to constrain the brain's mode of processing in waking, is turned down in NREM and is turned off in REM.

Chapter 4

Caravaggio
Running the Pons Motor Pattern Generator

A colleague calls to suggest that we go to Cambridge to see the Caravaggio paintings at the Fogg instead of taking a weekend—as planned—in Washington. I assent and find myself on a bicycle headed for Cambridge.

Suddenly, I feel resentful. A weekend in Washington would have been quite pleasant. How, and why, did my colleague change plans so arbitrarily and unilaterally?

The bicycle is moving along a narrow path on top of a steep grassy slope, in a quadrangle composed of Georgian college buildings. There is a tendency to ride down the bank, but the best route is at the crest.

After navigating the bank, I (now alone) reach the corner of the quad and try to find a way out—over rooftops and walls—while keeping the high level of the bank. There is no way out.

I think, "Damn, I'll have to go all the way back," implying to Brookline.

But, on turning back into the quad proper, I notice a path—indeed, an elaborate walk—leading out of the quad.

I marvel at the industry and meticulousness of the Harvard buildings and grounds. They have laid an exquisite walk of cut granite, not poured concrete.

As I round the corner of the walk, I see a workman and the job supervisor checking the work. Both wear white plastic hoods that cover their heads and shoulders. I assume that these hoods are required by OSHA to protect workers from toxic industrial fumes. The cement between the stones is not quite dry.

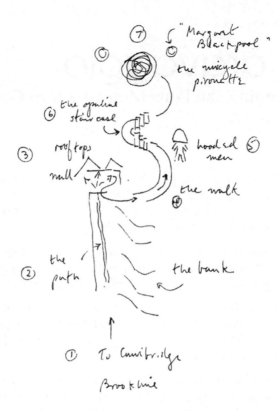

I ride past and reach the top of an opaline stairway. Each slab is transparent aqua lucite with white streaking visible in 3-D, like marble but clear. Again, I marvel at Harvard's attention to detail, but this time I also hear a student reading a course description: "The best thing about Byzantine Art 120 is the architectural approach afforded by a specially installed opaline stairway..." and so forth.

I think, "What lengths they will go to to please their students!"

But I also wonder how to navigate the stairs, which form an S, on a bicycle. Two solutions emerge: 1. Each step is a long diagonal ramp down with a stopping step at its end, and 2. the bicycle becomes a unicycle.

As a consequence, I let gravity carry me down, and with my gentle braking aided by the stops, I get down easily.

In the flat area below, I encounter two women, also on uni-cycles. They are probably social workers. One says, "Do you know Margaret Blackpool?" I reply, "Of course," convinced I have heard the name, fairly sure she is a social worker, but unable to recall what she looks like.

I then notice that by rapidly reversing my pedaling direction I can execute a spectacular set of curved trajectories or squiggles. My two female companions look on in admiration, saying, "You're quite good at that."

I demur with the statement "I've never done it before."

This dream occurred on a Tuesday night, but the recall was so strong that I could produce the preceding report on the following Thursday, two days later. In order to capture this dream memory, I lay in bed after waking, which occurred between 6:00 and 7:00 in the morning. I then rehearsed the details of the dream in my waking mind.

Since so many of my remembered dreams were harvested in the morning hours, it is important to realize that they cannot be considered typical of all my dreams. Would the same animation, bizarreness, and emotions prevail in dreams that occurred earlier in the night? Probably not. In fact, recent work by Roar Fosse suggests that the hallucinatory power of dreams increases across the night, while thinking declines early in the night and stays low thereafter. Highly hallucinatory dreams thus are most typical of early-morning REM mentation.

All the dreams in the preceding chapters were recalled after spontaneous awakenings. Would they have been equally vivid and animated if the awakenings were instrumental? We cannot know. This question deserves an answer that only new research can provide. That we may be in for a surprise is suggested by Fosse's finding that the emotion profile of instrumentally elicited dream reports contains much more positive emotion than spontaneously recalled dreams. It seems likely that negative emotions of some anxiety and anger dreams actually lead to awakening. Hence we are biased, by our own experience, to think of dream emotion as primarily negative.

Telephone conversation is an unusual dream behavior for me. Bicycle riding and other forms of self-propelled movement are much more common. The "colleague" who called me is unidentified. This is odd, especially since the suggestion that we visit an art museum instead of going to Washington suggests a person well known to me. In fact, I can't think of any of my professional acquaintances who would make such a suggestion. However, two people in my life at that time could have been fused in this scenario and together invited me to an art museum: Frank Sulloway and Margaret Warner, both historians of science. In waking life, I would have shown more anger toward anyone who changed travel plans as summarily as my unidentified dream telephonist did.

Off we go on the dream bike ride! Bike movement persists to the end of the dream with only one modification, when my bicycle becomes a unicycle when I encounter Margaret Blackpool in the last scene. To visualize this remarkable dream trajectory, consult the drawing, which shows the route and numbers the dream plot events along it.

The riding goes from easy and pleasant (if slightly hazardous [along the bank]) (1 and 2) to impossible [entrapment behind buildings] (3) to magical liberation via the granite walk (4) to improbable or impossible on the opaline stairway (6) and finally to the flat area (7) of the unicycles. This movement course with all its discontinuities and incongruities is typical of my dreams, which are related, as in this one, to improbable or impossible architectural pastiches. If I were cycling to Cambridge, as I have done many times in my waking life, it would be nothing like this. So my brain mind must be undergoing a continuous challenge to create the visual images and the associated cognitions that need to be integrated with my continuous sense of movement and my strong concept of destination. In other words, this is not a replay of waking experience. It is a novel scenario created by my dreaming brain. Where does it come from?

Working with Helene Porte, who was then a postdoctoral fellow in my lab, I set out to characterize dream movement with the same determination that I had attacked dream bizarreness. Bob

McCarley had previously noted that every sentence of every REM sleep dream report contained an action verb. That meant that dream consciousness was powerfully and persistently animated. And it is true, isn't it? We are always moving in our dreams. This obvious fact is deceptively simple. No one had ever investigated it before. Freudians and ex-Freudians were too busy interpreting dream content as either a derivation of unconscious drives or an expression of wake-state activity to see this surprisingly strong and constant feature of dreams.

Many of us, including me and many of the lab's experimental subjects, tend to be sedentary in our waking lives. One of our subjects, the dream journalist who was known as the Engine Man because of his fascination with locomotives, was actually a professional entomologist at the Smithsonian. Even though he sat at his desk in an office all day, he never dreamed about doing so. And I, who sat in my office with patients when I was not tuning in on nerve cells, never dreamed about that part of my life either. So much for the theory that dreams reflect waking experience. Rather, they simulate the movements of waking, and they do *not* simulate the far more usual periods of inactivity. I never dream of sitting at a table, writing and staring off into space. In other writings and in this chapter, I discuss the obvious implication of this fact for dream theory. In REM sleep especially, but to a significant extent in NREM sleep too, the brain must be running its motor programs.

Single-cell recording experiments, first championed by my NIMH mentor, Edward Evarts, have shown that the neurones of the motor cortex that are essential to intentional movement fire in clusters during REM. So do brain cells in other movement-supporting regions, including many subcortical nuclei of the anterior brain, the more posterior cerebellum, and the elements of Lorente's vestibulo-ocular reflex, the pontine giant cells, the vestibular nucleus cells, and the brain stem neurones that directly execute REM.

The brain is built to move and see. The drive to move and see, not unconscious wishes, is what constitutes the real latent content of dreams. Procedural memory tightens the binding of experience to movement. I can probably still ride a bike even though my poor balance and motor weakness might make the endeavor hazardous. How do I "remember?" I practice bike riding in my dreams! That is to say, my dreaming brain runs its motor programs, offline, in the safety of my bed.

In my Caravaggio dream, I can ride along crests and down stairs, and I am unfazed when my standard bike becomes a unicycle. Call it wish fulfillment, but this wish is fully conscious and fully pleasurable. The Engine Man may have been sedentary at his workplace, but he loved sports and described and drew many of his exotic dream trajectories. He golfed, played tennis, ran and walked, drove his car round and round in circles, rode on a flying carpet, and observed flying objects in his dreams. When I was doing the dream lucidity experiments on myself, I also flew through the air, eventually achieving a weightless, effortless soaring that was thoroughly delightful. Dreams of flying are not uncommon, and all spring from the same cause—our brain-minds are designed to envisage even movements that are physically impossible.

The architectural transmogrification of my Caravaggio dream is impressive: Georgian college buildings give way to rooftops and walls, to a granite walk, and finally to the opaline stairway. My rationalization about the unlikely turn of events—that Harvard will do anything for its students—is an ad hoc explanation typical of dreams. Harvard *is* good at developing and maintaining its architecture, but the elaborate details of the work are too much even for my dreaming brain-mind to accept. I attribute to a student passerby the marvelous course description of Byzantine Art 120, which the opaline stairway illustrates. Need I say that my waking mind knows of no architectural feature on the Harvard campus that has been created to illustrate course material?

When we were pouring over dream reports during that five-year search for the essence of dream bizarreness, we considered not only dream plot features but also the thoughts that the discontinuity and incongruity of plot features seemed to inspire in the dreamer. The category that was most frequently fulfilled by faulty dream thinking was called the ad hoc explanation to indicate that some degree of thinking had been necessitated by the dreaming brain's notice of bizarreness. More often than not, these ad hoc explanations were as illogical as the perceptual events they were supposed to explain. We then suggested the dreaming brain was throwing good cognitive money after bad. The English poet and dramatist John Dryden put it nicely in 1700, the year of his death:

Dreams are but interludes which fancy makes.

When Monarch Reason sleep, this mimic wakes!

Compounds a medley of disjointed things.

A mob of cobblers and a court of kings.

Our dreaming minds mimic the waking mind's habit of explaining why things are the way they are. The explanations are as bizarre as the events of the dream world. The point is that the brain has an intrinsic organization that guides it, experiences movements, and comes up with explanations for them.

My Caravaggio dream exemplifies the phenomenon of ad hoc explanation. In the lab, at first we included these mental phenomena within the category of dream bizarreness. Now we would separate the conceptual from the perceptual but still assume that they are intrinsically bound up with one another and that both suffer from a common defect—namely, the loss of executive functions due to the inactivation or underactivation of the frontal lobes in sleep.

In my Caravaggio dream, I try to explain the wondrous architecture I behold: the granite stone wall (possible and, at Harvard, even probable) and the opaline stairway with its aqua lucite risers (clearly impossible, even at Harvard) in terms of its connection to a course called Byzantine Art 120 (as if that explained anything). Even the Byzantines stopped short of aqua lucite stair steps! Yet this explanation seems to satisfy my dreaming brain's need for explanation. That's what I mean by "ad hoc."

Having gone off in search of my colleague at the Caravaggio exhibit at the Fogg and having negotiated the tricky dream stairway, I encounter two women who are "probably social workers." Why social workers? And why are social workers riding unicycles? I have no idea! My dreaming mind didn't know them any more then than my waking mind does now! And I really have no idea who Margaret Blackpool is, even though I say, "Of course I know her." Since they address me as if I were their colleague, I oblige by compliant lying. This is confabulation, pure and simple.

The failure to identify dream characters with confidence is one of the key formal aspects that underlies dream bizarreness. The failure to identify characters, to accept inadequate evidence for their presence, their function, or their identity, is an outcome of the impairment of cortical executive mechanisms. I don't even seem to notice that I have not, in fact, achieved my destination, which was the Caravaggio exhibit and the Fogg. Nor have I found my colleague, who was a "he," not two "shes."

In the previous chapter's Lobster dream, I recognized my colleagues Ralph Lydic and Barry Peterson. In this one, I have no idea whom I am going to meet in Cambridge, no idea who canceled our trip to Washington, and no real idea of who the two social workers were. Dream disorientation can be personal as well as temporal, spatial, and functional.

Notice that I even succumb to dream social pressure when asked, "Do you know Margaret Blackpool?" I reply, "Of course." Even though this is an outright lie, in the dream, it is sincere. And it works for me.

Confabulation works well for the patient suffering from Korsakov's psychosis, who has burned out his memory system because of excess drinking. Such patients often exhibit a garrulous charm, unaware that much of their discourse is just as fabricated as that of our dreams. We speculate that dream confabulation might result from the brain-mind's intrinsic capacity to create coherent narrative even when huge chunks of orientational data are missing. Just as in Korsakov's psychosis, the normal brain in REM has altered its memory capacities. Among the potential sources of dream amnesia-confabulation is the well-established disconnection of the hippocampus, essential to narrative memory construction, from the cortex, where the elements of such construction are stored.

In dreams the mental functions or dysfunctions of hallucination, disorientation, memory loss, and confabulation are common. These four mental state features constitute the organic mental syndrome. A syndrome is a collection of symptoms that, when seen together in a patient, indicate to the examining clinician a probable cause. Thus the syndrome of recent memory loss, disorientation, confabulation, and visual hallucination tells the physician that his patient is an alcoholic or drug addict with organic disease of the brain regions necessary to normal mental functions. In REM sleep, these same brain systems become dysfunctional. This temporary dysfunction is reversed upon awakening, indicating that it is state-dependent. When seen in acutely confused patients, the syndrome constitutes delirium. When seen in chronic patients, it suggests dementia. In both acute and chronic manifestations, the brain is weakened by the toxic effects of a drug, most commonly alcohol. Alcohol actively suppresses REM, and the delirium tremens (called rum shakes by some patients) that follows alcohol withdrawal is associated with monumental REM rebound. It is in this condition that we can truly say: Let the dreamer awaken, and you will see psychosis.

It is ironic that this dream-as-psychosis epigram is attributed to the psychoanalyst Carl Jung. In his early work with Manford Bleuler, his chief at the Burghösli Clinic in Zurich, Jung evinced the careful curiosity of a clinician-scientist. But when he threw in his lot with Sigmund Freud, the great Viennese conquistador, he seemed to forget the value of systematic observation. That no one else noticed the similarity of dreaming to the organic mental syndrome is testimony to the thrall that psychoanalysis cast upon psychiatry for almost a century.

The physiological basis of the delirium of dreaming is that certain regions of the frontal lobe become inactive as the brain shuts off its own supply of the chemicals used in waking to assure normal mental processes. Perhaps these two apparently separate causes are one. Nevertheless, the physiological basis of delirium in the acute organic mental syndrome is the same: intense rebound of REM sleep following its prolonged suppression. We speculate that it is the power of REM physiology, now heightened by the effects of prolonged suppression, that breaks the barrier into waking and causes the symptoms. This is dynamic and disastrous dissociation. In chronic organic brain syndrome, the damage to the midbrain and thalamic structures is permanent and irreversible. I hypothesize that the brain of such a patient is no longer capable of maintaining normal waking mental processes because it is in a hybrid state: many of the features of REM are mixed with those of waking.

In analyzing my Caravaggio dream, I note that I am conflicted about two proposals. Although the proposer in each case is unidentified, I wrote in the margin of my journal that I had lunch with Frank Sulloway at the Harvest Restaurant in Cambridge the next day. I might have cross-referenced him with the Caravaggio show, but I don't know if such a show actually existed. What I do remember is that the modernist painter Frank Stella had devoted his Norton Lectures to a discussion of Caravaggio, and I wanted to go because I had heard a bit about Caravaggio but did not then understand his popularity. This dream is related to my long-standing attempt to understand Italian culture.

I was getting to know Sulloway during this period because we had similarly critical views of Freud. The nice thing about our collegial friendship was that we had such different styles, techniques, and knowledge that we could learn from each other while sharing a deep interest in the same subject, the science of the mind. At the time of this dream, I was writing *The Dreaming Brain*, and Frank had just published *Freud, Biologist of the Mind*.

Sulloway could not have been the author of an invitation to Washington, but Margaret Warner, his younger historian of science colleague, could have been. Margaret was a medical student whom I recruited to the lab through the William James Seminars. During our time together, we went to Washington together once and had a wonderful time. The *Margaret* Blackpool social worker of the unicycle scene gives credence to my weekend-in-Washington hypothesis, but I do not suppose that the dream character, Margaret Blackpool, was a disguised stand-in for the other M.

A formal analysis such as the one I have given here does not identify either the Cambridge luncheon or the Washington weekend collaborators. There is no need to account for their muddy or obscure identities. After all, I *knew* who these people were, and I knew that my relationships with them were passionate, important, and problematic in many different ways. In other words, I did not need a murky dream to help me fend off recognizing the conflicts associated with these people. In the dream, they serve a purpose, but it is a relatively weak one compared to my sense of movement and my feelings of fear, awe, and fascination during the hallucinatory scenes of my bicycle trip. Speculation about the possible reasons for character misidentification or unidentification could have the unfortunate effect of obscuring this important point: We generally fail to access episodic memory during dreaming. Somewhere in my dreaming brain, all the rich data about Frank Sulloway and Margaret Warner lay intact but inaccessible to my dream consciousness. As soon as I woke up, I had no problem bringing this data to mind. So something about REM sleep might disable the retrieval of memory.

After our intensive study of dream bizarreness, we turned our attention to dream emotion. At first, we thought we would catch it in our bizarreness dragnet but, as I have emphasized, dream emotion was always congruent with dream plots even when the plots were full of incongruity and discontinuity. If a dream is anxiety-ridden, the plot reflects this feeling in a variety of ways. This is seen in the opening scene of my Lobster dream when my orientational confusion is associated with what I call frustration. If a dream is ecstatic, as the second part of Lobster and most of Caravaggio demonstrate, the plot consistently mirrors that effect even as it jumps from item to item or place to place.

We were surprised to find that the dream emotion spectrum was radically shifted to what we call the hot side. Thus, whether the dream reports were obtained after spontaneous or instrumental awakenings, three emotions were most prevalent:

- Joy/Elation
- Fear/Anxiety
- Anger

There was no evident difference between men and women in this range of emotions, and both sexes had surprisingly few cool emotions such as sadness, shame, or guilt. The direct expression of erotic feeling was likewise surprisingly low.

In my Caravaggio dream, I am elated, enjoying the ride and my mastery of the bike even when it seems dubious that I will be able to continue. Anxiety creeps in from time to time but is crowded out by successful action. Anger (or at least irritation) is also present, but it too takes a backseat to the pleasure of movement.

Now that we know the limbic regions of the temporal and frontal brain are selectively activated in REM sleep, we look forward to more detailed studies that attempt to associate specific emotions with specific areas of this large and complex brain system. To help this occur, we look beyond PET to fMRI because that technique will enable us to image online, in real time, while subjects dream, or do any number of other activities, in the scanner.

It is important to stress again the important advantages of MRI over PET and to explain why it has been difficult to use MRI in sleep studies. With PET imaging, the investigator gets one and only one look at the brain's regional activation pattern in each study. And the subject must be exposed to a radioactive isotope to yield that single image. This means that all the marvelous studies of sleep using PET are only "snapshots," not movies and not even time-lapse photographs.

While both PET and MRI both have limited degrees of spatial resolution, MRI's temporal resolution far exceeds that of PET. With MRI, a continuous succession of images can be collected across the entire night of sleep. This allows second-to-second comparisons of activation to be made. That's the good news. The bad news is that MRI depends on the frequent induction of magnetic field changes. To achieve these changes, the magnet literally clanks, and the noise is as disrupting to sleep as the fields are disruptive to the electrographic recordings that are used to objectify the brain states.

The Caravaggio dream reveals my appreciation of architecture. I love complex and beautiful building forms, especially when they are successfully combined with the landscape, as they are in this dream. I enjoy my own amateur efforts with architectural design. Of course, the experiments at my farm in Vermont and at my house in Brookline have been much more modest than these dream images, but as Browning said, "A man's reach should exceed his grasp, or what's a heaven for?"

It may not be a coincidence that on the very day of this dream, a Vermont mason named Karl Armstrong was building a patio outside my kitchen porch in Vermont. He was to use the granite sills I had scrounged from my neighbor, Steve Cahill, to frame the patio. The granite sill frame was to be filled and leveled with gravel and finally topped with flat stones my sons and I had hauled from a brook bed outside Burke Hollow. I was worried about whether Karl could follow my instructions and whether he would use mortar.

But the dream is not ostensibly about Vermont. On its face, it doesn't seem to have anything to do with Vermont. Vermont workers often show little attention to occupational hazards. They would *never* wear protective clothing or plastic hoods. My farmer neighbor, Marshall Newland, breathed hay dust four to six hours a day and suffered from asthma all his life before dying of farmer's lung. His brother-in-law Wesley Ward slipped when he stepped in the fresh paint he had just put on my barn roof and slid to the edge before stopping himself just short of a 20-foot fall. Later, when going to fetch some paint in another barn, he fell through the floor, broke his back, and was out of work for six months. When his wife came to tell me all this, she apologized profusely for the inconvenience to me and promised that he would finish the job when he got better! "It's just one of those things that happens," she said. No lawsuit, no workmen's compensation, no protective hood, and not even a rope to stop Wesley from sliding down the roof? I can't help thinking that the OSHA guys in my dream have something to do with Vermont occupational hazards.

In the Caravaggio dream I attribute these splendid dream buildings, stairways, and walks to Harvard. While I do admire Harvard's architecture—especially on the maintenance side—the dream creations are mine! I can't realize that fact because I am caught by their hallucinatory intensity. I have lost self-reflective awareness. Here, again, "seeing is believing!"

Leonardo da Vinci is said to have asked: Why does the eye see such a thing more clearly in dreams than when awake? The brain scientist's answer is that the visual brain, which is what Leonardo means by the eye, is more intensely activated in sleep than in waking, and all this activation is internal, because no light-derived information arises from the retina. The data from basic sleep research, showing dramatic bursts of neuronal firing in the eye movement control centers of the brain stem and in the visual processing centers of the forebrain, together with PET scan data of the human brain in REM, form the evidential basis of this claim.

But the pharmacological data stands on its own in testimony to the brain's remarkably convincing simulation of visual reality in the absence of any visual stimulation from the outside world. So Leonardo's question, asked in the 15th century, presages several important implications of modern sleep research.

One is the fact that all vision depends on internal as well as external form. This means that the form of vision is shaped, in part, by the brain itself. A second point is that, when sufficiently activated and suitably modulated, the brain can see quite well on its own. Third, this realization is of interest not only to dream theory but to our way of thinking about all internal visual processes.

Consider the ability to visualize while our eyes are open and we are awake. I see the volcano on the island of Stromboli spewing red-hot lava streams even as I stare across the placid blue sea stretching away from me as I write. Mental imagery might depend on our capacity to activate, at will, the visual centers of our brains.

Some of us can see objects wholly integrated in their visual scenes with convincing clarity. In the case of highly hypnotizable subjects, this capacity is called visual hallucinosis. Such skill borders on the pathological even if it is at an extreme of the normal. I say this because even "normals" like me can be induced to visually hallucinate if sleep-deprived. When I was doing all-night sleep recordings at NIMH in the early 1960s, I regularly welcomed entirely hallucinated visitors who arrived at the door of my lab between 4:00 and 6:00 a.m.

In this discussion, we are in the well-marked territory of William James, whose 1912 masterpiece, *Varieties of Religious Experience*, anticipated much that we have since discovered. The fine line between the normal and the pathological capacity to hallucinate visually is difficult to discern, but it is clear that suddenly abstaining from high doses of REM-suppressant drugs, such as alcohol and amphetamine, might unleash a storm of formed visual images—all of which are inserted into the waking visual scene. As noted, visual hallucinosis, while common in organic psychosis, is rather unusual in "functional" illnesses such as schizophrenia and affective psychoses where auditory hallucinations predominate.

When I was an inexperienced trainee psychiatrist, I thought I could talk my schizophrenia patients out of their hallucinations. But I soon found that I was no more successful in forcing this insight than I was at noticing that I myself was dreaming when I was asleep. In spite of brief and fleeting episodes of lucidity during my early experimental life, I am constantly fooled, in my dreams, into thinking I am awake.

My love of bikes and architecture are the main building blocks of this dream about my life at Harvard, where my enjoyment of collegiality had recently increased. I used my research and teaching interests to create relationships, and I used my physical presence at Harvard to stimulate and reinforce the architect in me. All these themes are revealed, along with my feelings about them: There is frustration (about choices and directions), joy (about movement), anxiety (about achievement), wonderment (about landscapes and buildings), and even comic exhibitionism. Everything in our dreams is based on experiences we have had and identities we have come across in our waking lives. The emotions that are tied to those identities are revealed in dreams, especially the hot emotions. But the dreams are *de novo* creations of the sleeping brain.

Chapter 5

Italian Romance

Playing Hide-and-Seek with Dopamine

I have met and established a sexual rapport with a beautiful Italian woman named Francesca Vivaldi. We have an unspoken but clearly understood rendezvous in her hotel room (or is it a boat cabin?).

The décor of the corridor is ultra-modern (as in the Italy of Milano), and as I round a corner, I note that the wall is a continuous piece of poured white plastic, as is common now in single-piece bathroom installations. In fact, the door of one of the rooms is a shower, and I see a young boy, naked, hurrying in to avoid embarrassment. I am relieved because I too am furtive in my quest. It seems that Francesca's whole family may be in this section of whatever place this is. I realize that I could ask directions to her room (I don't have the number), but that might reveal my aim.

I round another corner and come to an open salon, where I see my mother, her sisters, and other women putting on elaborate foundation garments (bras, girdles, pantyhose), again quite pale white, smooth, and surreal. They giggle at their undress but are, at the same time, matter-of-fact, as if in preparation for an important family event, like a wedding.

My father, fully dressed, is serving milk punch and offers me a large glass. "Always the impresario-entertainer" I think to myself, and compliment him: "Fabulous milk punch!" He says, or indicates by his thoughts, "Yes, George made it." And my cousin, George, is indeed standing by. I am recruited to help refill the cut-glass punch bowl by carrying it to the kitchen. It is unduly heavy, and I jokingly strain to lift it, thinking, "He thinks I'm going to drop it."

I was sleeping at home in Brookline with my first wife, Joan, and I have vivid recall of the recorded episode. As is typical of my early morning awakenings, I was also sure that the remembered episode followed at least 10 minutes of earlier dreaming that I could not recall. The fact that I had an erection indicates that this dream was driven by REM sleep brain activation, because erection and clitoral engorgement in females are constant features of REM sleep physiology whether or not the associated dream content is sexual. In keeping with this supposition, the remembered content is bizarre in its detail, has a scene shift in the middle of its course, and is characterized by strong and unifying emotion throughout. In scene 1, I am seeking and lustful; in scene 2, I am humorously amorous with my parents.

When I woke up, I told this dream to Joan, with whom I was hoping to have a complete reconciliation and a truly open marriage. She asked me if I was having an affair in Italy, a remark that indicated her unease at this prospect. If I were not a scientist, I might even think that this dream was prophetic. While it was true that I had not yet had an affair in Italy, I was trying hard to have one. More important is the fact that after the breakup of my marriage in late 1992, it was to Italy that I went, and it was in Italy where I met Lia Silvestri, the woman who became my second wife.

My white plastic dream is typically vivid, especially in the visuomotor and emotional domains. With the exception of "Francesca Vivaldi" (the object of my quest, who does not appear), I see all the other dream characters clearly and note with interest the details of their dress (or undress). The persistence of smoothness and whiteness across all three scenes (the single-piece bathroom wall in scene 1, the elaborate foundation garments in scene 2, and the milk punch in scene 3) is unusual and suggests a formal associational rule, color, that also unifies the content.

The main point of this part of the discussion is to address the intensity and "surreal" clarity of dream vision. In calling such intense internal vision hallucinatory, I am not suggesting that dreaming is formally identical to delirium—only that it has all the formal properties of that abnormal state. Delirium is one form of mental illness that is caused, like REM sleep, by shifts in brain chemistry.

So vivid is my dream vision (and so weak is my self-reflective awareness) that I assume I am awake and living out the unbelievable dream scene. This credulous quality of dream thinking is what leads me to characterize dreaming as delusional. Had I access to either my memory or my critical faculties, I would have realized that Francesca Vivaldi was a figment of my imagination, that my mother and her sisters are all either dead or totally infirm, and that my father was well beyond the milk punch stage of his development. But, like the dreams of my blind Iranian patient, who saw his postman father in his dreams, lost and even dead family members can appear, in all their glory, in our dream vision.

In my dreams, it is typical to be unsure of where I am. I am spatially disoriented. Is this a hotel or a boat? And the door of a room becomes a shower into which a young boy, unidentified except by his nakedness, suddenly runs. The fear of discovery mounts in me. I know I am doing something naughty, and I want to avoid detection by Francesca's family, who may be staying in this section (cognitive uncertainty). As it turns out, I don't even know where I am going (more spatial disorientation), and I rationalize not asking directions (ad hoc explanation) on the basis of discretion. As we all know by now, there is a much simpler explanation for the bizarreness of dreams. My orientational instability is a part of functional delirium caused by the shift in balance of brain neuromodulators.

The only thing that is clear is my motive—sex! My seductive mission is not concealed at all. It is as naked as the young boy darting into the shower. Whenever I have sexual or seduction dreams, they are always as transparent as this one is. But sex and seduction are no more common in my dreams than are such fantasies and behavior in my waking life. If disguise and censorship are trying to put a psychological fig leaf on my id, they aren't covering it very well!

Is erotic quest properly defined as an emotion? We have said "yes" in our published articles, and in this dream it is certainly as much a motive, or a driving force, as it is a feeling state. Francesca Vivaldi is never found, and the appearance of my parents changes my escapade into a more civilized search for a bride, whom they would accept. This emotional evolution from sexual predation through marriage preparation to celebration is a sop to convention that I do not always observe when awake. In addition to eros, in scene 2 I experienced the emotions of surprise, pleasure, and humor, which far outweigh the anxiety of scene 1.

As noted in discussing dream emotion, erotic feelings are quite rare in dreams. This fact is difficult to explain given the physiology of REM. One would think that the almost-constant erection in men and clitoral engorgement in women signals strong, centrally driven sexual excitement. But this is not the case, and that's bad news for several important theories of the brain-mind.

The first victim is, of course, psychoanalysis. For Freud, Charcôt's dictum "Toujours la chose genitale" was repeatedly echoed. But there is little evidence for a role of sexually driven dreams in modern sleep research, and even die-hard Freudians have turned away from this emphasis on sexuality.

The James-Lange Theory of emotion also dies on the sword of the observed dissociation of peripheral sexual excitement (erection, and so forth.) and central emotional expression (anxiety, anger, and elation, but not much sex). Here, our hero, William James, makes a colossal mistake when he assumes that our felt emotion is only a reading of our bodily sensations. In REM sleep, at least, nothing could be further from the truth.

Finally, the paucity of sex in dreams is disappointing to the hedonist seeking imaginary satisfaction in his dreams. Sexual dreams do sometimes occur, of course, and they can be fully consummatory, as in the so-called "wet" dreams of adolescence and early adult life. But when they occur later in life, they are more commonly dry, and the older dreamer often wakes up thrusting uselessly towards the elusive orgasm. Lucid dream training may be an antidote to the sexual frustration of dreaming, but even in this case, it is no substitute for the real thing.

Is there a compromise between the traditional psychoanalytic view of dreaming as wish fulfillment and modern dream science? Perhaps. If one gives up the disguise-censorship notion, as Mark Solms has done, it remains to explain the questing oniric behavior exemplified by my search for Francesca Vivaldi.

In fact, most of my dreams are not only animated but motivated, as if the investigative, experimental, and sybaritic parts of my brain-mind were turned on in sleep. In Dreamstage with

Elephants, I am trying to solve display problems. In Lobster Brain, I am on a culinary quest. In Caravaggio, I want to admire architecture from a bicycle. And here, in Francesca Vivaldi, I am looking for the perfect woman.

But, and it is a large *but*, my dream questing is not disguised. Rather, it is transparent. Strong, but transparent. If these dreams reveal unconscious wishes, something eludes me. All the motives for my Italian Romance dream are conscious. I agree that their disparate parts are held together by the glue of motivation. But, I ask, what else is new? Can you imagine activating the brain and not seeing organization at some level? I can't.

Psychoanalysts, like Mark Solms, are scrambling to get back in the game after the game is all but over. They say, for instance, that REM sleep is generated not only by acetylcholine, as we have shown, but also by dopamine—the motivation- and movement-initiating neuromodulator. It is true that acetylcholine and dopamine are comodulators of the motor and brain reward systems. But so far, attempts to show differences in dopamine release between waking and REM (and NREM, for that matter) have failed. If it is shown that dopamine release is increased in REM, we will welcome that finding as much as the psychoanalysts will. It remains possible, however, that in REM sleep dopamine is more potent than it is in waking, because it is not competing with norepinephrine and serotonin. In my waking life, as in my dreams, I am always questing: for wine, for women, and for adventure. But in waking, there is something distinctive about my questing that my dreaming entirely lacks. It is the power of thought, which serves to guide and restrain the complex engine that is my brain-mind.

This Italian Romance dream roots my sensuality and sociability in my early family life. It is one of a very few of my recorded dreams in which both my parents appear. Note how seamlessly they are integrated into my increasingly romantic European world. I have been going to Italy more and more since my 1984 visit with Francine Fonta. All my French friends said that if I loved France, I would love Italy even more, as they themselves do! The French regard Italy the way I once regarded France: as a Romantic culture in which to more deeply experience the sensuality of life.

In 1985 I had not found anyone like Francesca Vivaldi. Italian women played much harder to get than the French, who seemed to jump into my arms. Of course, I spoke French fluently and Italian only haltingly, but that is not the reason for the relative reticence of Italian women. Italian women are more reserved because they are more church- and family-oriented than are the French.

During earlier trips to Italy, as in 1977 when I went to Riolo to begin writing *The Dreaming Brain*, I had gone through Rome. My colleague, Mario Bertini, introduced me to Cristiano Violani and later to Fabrizio Doricchi, who became my guides, translators, and promoters. Cristiano's mother, Franca Violani, could well be the prototype for the Francesca Vivaldi of my dreams. Despite being ten years her junior, I was deeply attracted to Franca and had just received a letter from her. We often went together to her summerhouse in San Felice, where she prepared simple but marvelous Italian lunches featuring her own homemade antipasti. We sat in the shade and talked and talked. Why should she be called Vivaldi in my dream? Vivaldi is as generic an Italian name as one can imagine. I also know several Vivaldis, since Ennio Vivaldi, a Chilean scientist of Italian extraction, worked in my lab around 1980.

The dream setting doesn't fit with the real-life Franca Violani either. She would have an antique habitat, not the ultra-modern décor of Milano, and certainly not white plastic. Milano, however, is one of my first points of contact with Italy and with Italian women. I had met Fedora Smirne in 1975 at the International Congress of Sleep Research in Edinburgh and developed a very poetic, comical, and quasi-Romantic friendship with her. We wrote limericks about famous sleep scientists, which I read at International Sleep Research banquets. I can still remember two of them. Here's one:

Eugene Aserinsky's piacere	(pleasure)
Was his female assistant's sedere	(ass)
When he wiggled his occhio	(eyes)
She cried out "Finocchio"	(fairy, gay)
And that made them both feel godere	(pleasure)

Working in Nathaniel Kleitman's lab in 1953, Eugene Aserinsky discovered REM sleep. But as a no-nonsense physiologist, he was diffident about the connection of REM with dreaming. Aserinsky's unique character was recently vividly sketched by Chip Brown in the *Smithsonian* magazine. In 1957, Bill Dement picked up the dream theme and ushered in the sleep and dream lab era, which was just beginning to wind down in 1975. No one would ever attribute sexual shenanigans to Eugene Aserinsky. No one but me, in a dream!

Another limerick was about my psychoanalytically dedicated colleague, Milt Kramer, then at the University of Cincinnati:

My dream of Milt Kramer's bris	(circumcision)
Went a little something like this:	(cognitive uncertainty)
A choir of Castrati	(physiological sopranos)
From the Villa i Tatti	(in Florence)
Sang "Requiescat in Pace" Penis	(Rest in Peace)

There is little in this verse to signal Kramer's important contributions to dream research. Milt was, and remains, convinced that dreams have personal meanings. I always agreed with that. Milt also thought that psychoanalysis was the best existing theoretical framework for understanding dreams. This I deeply doubt. Perhaps I am punishing Milt for his Freudian adherence with this circumcision-castration poem. The words just came to me. I knew Milt was Jewish, so the bris made sense. All the rest I owe to Fedora and my own love of the Italian language.

What is art, and how do artists make it? My answer is that art is the emotionally salient representation of form. Artists make it by tapping into their feelings and imagery while they are awake. Sometimes, they tell us, they go to the dream world for inspiration. This is especially true of the surrealists, who were very much influenced by their French psychiatrist leader, André Breton. Breton was familiar with Freud's theories. It is also true of many more mainstream artists, such as Paul Klee, who asserted that creativity sprang from the same deep psychic forces as dreams, and René Magritte, who portrayed the unique formal properties of dreams in his paintings.

My point is that all dream-making is fundamentally artistic. This means that when we dream, we all experience something like artistic creation. We invent complex worlds, inhabit them, and enliven them with feeling but then, sadly, leave them behind. Does this mean that dream reports are works of art? Certainly not. If they are embellished, they may become literary, but in so doing, they may lose their value to science. Are dream reports scientific data, then? Yes, indeed. The first task of any science is scrupulous description. But reports are only reports. And they are never as richly inventive as the dreams themselves.

So it is crucial to distinguish between dreaming as proto-artistic and art as something made by waking brain-minds that speaks to us about the emotional salience of form. I am arguing that more sensitivity to the creative mode of even quite commonplace phenomena as dreams can help us better understand and respond to art.

There is another side to this story. It is the use of sleep and dreaming to advance scientific or artistic endeavors. There is no firm evidence on this question. I read such meager systematic evidence as has been collected as discouraging. But the anecdotes persist, and for me, personally, sleep does seem like a good way to incubate ideas, be they artistic, scientific, or even administrative. One example is the composition of occasional poetry, like my Edinburgh limericks.

My friend Duncan Nelson does this with great speed and effectiveness while fully awake. He collects data and then puts himself offline to the event and/or person he wants to portray. This enables him to assemble the facts he has collected, imbue them with passion, and express them in words well chosen to delight the listener by both rhyme and reason. I can't do that. But I can write my much-less-effective poems if I make a list of words associated with my topic and then go to sleep. When I wake up, the poem essentially writes itself, as did my Edinburgh limericks.

What these two little poems do is celebrate sexual voyeurism. I wish dreaming were better at simulating sex, so I attribute indiscreet motives to Eugene Aserinsky, the sober discoverer of REM sleep who turned his back on all the dream brouhaha that followed. I then go on to make fun of Freud's sexual theories by investing them in Milton Kramer, a strong defender of the psychoanalytic faith, and I mock his adherence to Freud by having a chorus of eunuchs sing his sex to rest.

My Italian Romance dream takes a surprising turn! As I round a corner, I encounter *my* family, not Francesca's, as I had consciously feared. Seeing my mother and other women, some of whom may be her sisters, putting on their underwear continues the theme of female-chasing but suddenly—and effectively— bowdlerizes it. If there is to be a wedding, it is socially acceptable to be undressed, and even my female relatives are comfortable with that! "In dreams begins responsibility," said the poet Carl Shapiro.

November 12th was the day before my mother's 79th birthday. Although she was already hopelessly demented due to Alzheimer's disease, I still felt determined to please her with emotionally salient gifts, like a happy dream wedding (instead of my ruined wake-state marriage). My mother was very conventional when it came to love, marriage, and sex. She was steadfastly faithful to her husband and children to the end. My father had a gleam in his eye but never strayed either, as far as I know. As my mother fell deeper and deeper into the dementia of Alzheimer's, my father went more and more often to visit Frau Jenni, his housekeeper and mountain-climbing companion in the Haslital Valley of the Swiss Alps above Interlaken. His children all hoped he might be having an affair with her, but that may well have been our imagination rather than his reality.

Under the surface of matronly decorum danced a playful, almost naughty-girl spirit in my mother. I remember, with great pleasure, her tomboy participation in our early baseball games. She was both permissive and exhibitionistic. My fascination with sexual anatomy began with playing doctor with Marilyn Phelps at age 5. I remember asking my mother, "If real doctors get paid for this, why doesn't everyone go into medicine?"

My mother knew about these early innocent games and about my later, more hazardous sexual experiments. But she never scolded me or even warned me about the consequences of my impulsive liaisons. She was also unconcerned about exposure of her own body. At age 6, as I sat in the bathtub, facing her with my feet between her legs, I was mystified by her pubic hair, which seemed to conceal her sex. Where was it, I wondered? The breasts were much easier to understand, especially when I saw her nursing my brother Bruce (when I was 10), but the structure of the female pudendum escaped my analysis until late adolescence.

My father was less directly communicative about these matters, which may explain why he was fully dressed in my Italian Romance dream. But he communicated his laissez-faire attitude quite clearly. It went along with the witty host charm that is exaggerated in my dream. I never saw him serve milk punch to a crowd, but he did like to make martinis, Manhattans, and even sidecars, which he served to his guests along with his little jokes and puns.

My father's reference to George as the dream punch-maker could be an homage to my cousin, George Chandler, who had a genuine host charm of his own. I knew George during my Washington, D.C. days (1961–1963), when we lived in the same apartment building at 3850 Tunlaw Road. In fact, it was childless George who drove my wife, Joan, to the Walter Reed Hospital to deliver our first child, Ian, while I was doing an all-night sleep recording in Bethesda.

My father's lack of confidence in me as punch bowl carrier in the dream is significant. Although he taught me many things and encouraged my work ethic in all my career domains, he was never able to reward me with praise. This may be why I am still nervous in the face of authority. My perverse reaction is to pretend that the dream punch bowl is heavy, hoping to make him more nervous about my dropping it. But when it came to sexual experimentation, my father was as noninterfering as my mother. "Don't ask, don't tell" could have been their motto.

Since this dream has an Italian lover theme, it could be related to my very intense and actively homosexual relationship with Salvatore Bruno, who was my baseball hero and lover from age 13 (7th grade) to 15 (9th grade). This affair flourished, hit the wall, and ended under my parents' noses without a word ever having been said about it, except that they too admired Sal! He was very special.

After Sal, I took up with my psychologist mentor, Page Sharp, who loved and nourished me through high school (ages 15–19). Known to his family and friends as Cupie (for Cupid, the God of Love), this wonderful man had developed his unique style as a volunteer in the French Medical Corps from 1915–1918 and through the early '20s as a banana plantation manager for the United Fruit Company in Central America. When he finally returned to the U.S., he went into post-graduate training in educational psychology with Samuel Orton at Columbia and later moved to Hartford, where he set up his practice in a lovely old house across the street from the Mark Twain mansion. Sharp was an intense Francophile as well as an expert in the diagnosis and treatment of dyslexia.

There are many points to be made of this, but let's keep the focus on sex. Europeans have a more liberal attitude about sex than Americans, but my parents, who were as American as apple pie, were tolerant of my adoption of very liberal sexual mores. At age 18, when I finally decided I had to break off with Cupie—on the eve of my going away to England and Europe in 1951—I consulted with my father about how to do it. He said, simply, "You'll find a way. Cupie is a very nice man." Here, amazingly, he was expressing a confidence in me that did not extend to dream punch bowls.

Freudians would, no doubt, be impressed with the oedipal structure of my dream. They might say that the sexual quest has nothing, really, to do with Italian lovers. This is just the latent content cover for my incestuous yearning for my mother, a yearning that my father effectively quashes by giving me the difficult task of transporting the "milk punch" bowl, a howling maternal symbol if there ever was one. The problem with this kind of hypothesis is that I am quite aware of my affection for my mother, and it is very poorly disguised in this dream. As for my father's role, it is more cooperative than contrary. He was always complicit with my experiments and would have been as happy as my mother with the outcome of my quest.

Two of Freud's lasting gifts to us are the value of free association and the concept of transference. Transference means that we treat people in our current life according to expectations learned earlier, often at the hands of parents. These two building blocks of psychodynamic psychotherapy are not invalidated by anything modern neuroscience has dredged up.

As for the issue of association, however, we should point out that the concept had been in the air since the early 19th century when David Hartley, the father of the British School of Associationism, first proposed that memory was categorically organized. Today, cognitive neuroscientists use the term *semantic network* to imply interconnections of words and meanings embedded (somehow) in interconnected neural networks.

I could associate endlessly with dreams like Francesca Vivaldi because it is populated by characters of great importance to me, including my parents. And because of the bizarre way in which they are presented, it is tempting to assume that my mind has resorted to symbolic or metaphoric transformations of these historically significant people. The bizarre representation of my mother and her sisters (in their underwear) and my father (as a milk-punch purveyor) could be seen as defensive transformations designed to disguise my wish to sleep with my mother and neuter my father.

The alternative offered by modern dream science is that, yes, these representations are symbolic and even metaphorical but, no, they are not defensive. On the contrary, they reveal, rather than conceal, details of my relationship with my parents that can be fruitfully explored via a discussion of my Francesca Vivaldi dream.

So far, so good for Freudian revisionism. Give up disguise-censorship but keep symbolism and the old dream theory doesn't look too bad. I honestly think that's what most psychodynamic psychotherapists would now say. But wait a minute. Why should my Francesca Vivaldi be a better place to start an exploration of my feelings toward my parents than any other point in my memory of them? The fact that the presentation is bizarre guarantees no special or privileged status to the dream images. I could arrive at the same points in the memory network that extends around my parents simply by recalling the events that my reading of the dream report brings to mind.

One answer might be that the waking mind, operating in its linear logical mode, is not as good as the dreaming mind in linking associations metaphorically. This fits with my assertion that dreaming is more autocreative than waking consciousness. But this proposition, a variant on Freud's "Royal Road" analogy, has never been critically tested. As far as I know, it has never even been questioned by psychoanalysts. Psychoanalysts tend to assume that because Freud said it, it must be true!

Having issued that caveat, I must confess that I enjoy talking about dreams with my family, my colleagues, and my patients. I view dreams as privileged communications from one part of myself (call it the unconscious if you will) to another (my waking consciousness). There can be no doubt that our dreams are trying to tell us something about how our memories are organized, that associationism is alive and well, and that dream discussions will continue no matter what we sleep scientists say!

For those who believe in precognition or in dreams as prophecy, Francesca Vivaldi is very "big" indeed. It foretells my destiny quite precisely. In 1989, something about me knew I would marry an Italian! But this is clear only in retrospect. To make such a theory stick, we need to make the predictions ahead of time, not retrospectively, and the ratio of hits to misses would need to exceed chance to support the precognitive hypothesis. I myself believe that there were indications, even in 1985, that I was headed for Italian romance. But there was nothing unconscious, mysterious, or unacceptable about that motive.

So where do I stand on the question of dream interpretation? Is it scientifically defensible or not? I am sure to disappoint most readers when I say that while the case is still open, all the evidence points against an affirmative answer. As time goes on, and more and more is learned about how the brain works in sleep, it seems to me that we will be forced to conclude that the chance of falling into error with any formulated approach to dream interpretation is still greater than the scientific benefits of taking the risk. When Freud said that he believed that one day, physics and chemistry would explain all psychic phenomena, including dreams, he was not counting on having his ideas replaced by science, but that seems to be exactly what is happening.

Does this mean that dreaming is a mere epiphenomenon? Does it perhaps mean that the conscious experience of the brain's self-activation in sleep is better forgotten? Maybe. But it does not mean, and activation-synthesis does not argue, that dreams make no sense at all or that dreaming should not be scientifically investigated.

REM sleep is highly conserved in mammals, and its functions are clearly vital to life. Yet dreaming as conscious experience might still be an epiphenomenon. As we focus more and more closely on the role that sleep plays in learning and memory, we are likely to appreciate dreaming as a clue to the reinforcement of remote associations, to the reorganization of memories, and to the regulation of the psychological self. But it is probably neither the guardian of sleep nor the Royal Road to the unconscious as Freud asserted.

As it turned out, it was in Italy that I found what I was looking for in my second wife, Lia Silvestri. Our liaison and marriage have been sensual, and we share a commitment to energetic, creative work and an equally strong commitment to marriage and family. Now that I am infirm and am more of a burden than a creative companion, I am grateful for the values that allow my wife to love me and care for me with unselfish dignity. I don't recall ever dreaming that would happen.

Chapter 6

Ed Evarts and Mickey Mantle

Scrambling the Hippocampus

Ed Evarts and Mickey Mantle are with me in a room where a grant review committee has recently met. The atmosphere is tense.

In one scene, Evarts (who died about a year ago) asks me, "Have you had a mood swing?" (implying that I am not my normal ebullient self). I say, "No, but I am bothered by this herpes sore" (on the left of my lips). Ed recoils in horror, saying, "Don't come near me unless you're having it treated!" This confuses me, because I am unaware of any effective treatment for herpes. But recognizing his authority—and not wishing to appear ignorant in case a treatment has recently been discovered—I say nothing.

Mickey Mantle (who is as stocky and handsome in my dream as he was in his ball-playing prime) looks a bit downcast. I try to cheer him up by inquiring about what's bothering him. He says, "Evarts says my grant request was papered over by the committee." I express surprise, saying that I had been one of three reviewers and had not found it weak. "So you reviewed it!" exclaims Mantle indignantly. He then begins to badger a secretary who is making unsuccessful arrangements for him to be picked up by helicopter. I counter weakly by pointing out that I was only one of three reviewers and that the decision was up to Evarts.

I then consider, but reject, the idea of inviting Mantle to stay for supper. It would surprise and delight my son, Ian, to meet him. But I hardly know Mantle, and he is irritable. Better to let him go.

Every year, from 1970 to 1990, I went with my first family to Louis and Katherine Kane's house in Ogunquit, Maine. On the night of my Ed Evarts and Mickey Mantle dream, my sleep was fitful because we had just enjoyed the classic July 4th dinner, which always featured a whole, huge salmon (grilled by Louis on his porch) and lots of wine. Because alcohol suppresses REM, I was probably acutely REM-deprived, and my late-morning sleep might have been propelled by the beginning of the REM rebound that typically follows even short-term REM deprivation. The brain's tendency to make up lost sleep has been recognized since time immemorial. A good way to ensure prompt and deep sleep is to abstain from it until the rebound pressure takes over.

In the early days of REM sleep science, it was assumed that if a subject was deprived of REM, he would also be deprived of dreaming, and that would be bad for the mind. Dream deprivation was a paradigm forged by a merger of the new sleep physiology and Freudian psychoanalysis. For its champions, William Dement and Charles Fisher, then working in New York, dreaming was a hydraulic-like escape valve for the psyche. If this valve were closed by REM deprivation, psychosis would very likely ensue. This hydraulic model was borrowed from Freud.

It was, of course, immoral to test this idea with human subjects. The working hypothesis was that dream deprivation would cause psychosis. Psychosis is unpleasant, even if it is quickly reversible. And no one could really know that it was reversible. Nonetheless, subjects were REM-deprived, and some of them did become psychotic. In a famous publicity stunt, the disc jockey Peter Tripp was sleep deprived in his broadcasting booth and became more audibly paranoid as his deprivation proceeded. The dream-deprivation paradigm had two other more-substantive problems:

- The assumption that all dreaming was associated with REM was wrong.
- The failure to include adequate controls via NREM awakenings was an oversight.

Both problems impaired the specificity of the results because no one could be sure they had really deprived the subjects of dreaming, and they could not conclude that REM sleep was more important to psychic equilibrium than NREM sleep.

One thing was clear from the early sleep deprivation experiments: The brain set a high store on sleep. If even 30 minutes of any stage of sleep were lost, there tended to be a payback.

Early dream theorists thought dreams were actually caused by indigestion of some other extracerebral source. Our scientific predecessors who were committed to the reflex doctrine thought the brain activation that underlay dreaming was a direct response to stimuli that caused partial arousal from sleep. We now know that the brain activation of REM sleep occurs on its own and is, if anything, impeded by the sort of indigestion that kept me awake that night in Maine.

My Evarts-Mantle dream was easily recalled because I was sleeping in on a weekend morning. Anyone wishing to increase dream recall will find that it is greatly enhanced by delaying getting up and by continuing to doze on weekend mornings. The sleep that occurs in these circumstances is either Stage I, REM, or very light NREM (which I call Stage I-and-a-half).

My dream was recorded promptly and in detail because I always take my journal with me when I travel. If, as in this case, I am forced to stay indoors because of bad weather, I have the time to focus my attention and write my journal entries right after breakfast.

This dream was high in visual hallucinatory perception. I could see Ed Evarts clearly and could describe Mickey Mantle in detail. They looked just the way they did in 1962! There was also auditory detail: The conversations are decidedly bizarre, but they proceed along one narrow thematic road—the grant review. Incongruity was rampant: Evarts and Mantle do not fit together in the grant review frame, so my brain is forced to give ad hoc explanations to accommodate Mantle.

In speaking to me about my mood and the risk of herpes, Evarts was not behaving in a characteristic manner. Remember, too, that Evarts had died before 1986! In the dream, I had no insight into or even puzzlement about this impossible detail. The two emotions that predominated are typical of dreams recollected upon spontaneous awakening: fear (mine of Evarts, Mantle's of the grant review process) and anger (Mantle is clearly vexed and "irritable"; Evarts's rejection of me is nearly hostile). Not so typical are sadness (Mantle looking downcast) and empathy (I was trying to cheer him up).

Speculating about its physiological provenance, it seems likely that this dream came from late-night NREM sleep. There was very little motoric action and no scene change, and the feelings were muted.

It is easy to see why I might associate Ed Evarts with the grant review process. I knew from the grapevine that Ed had reviewed and approved my first grant application to the NIH in 1968. This was five years after I had worked in his NIH lab (in 1963). I had decided to go to France in 1963 to work with Michel Jouvet with Ed's support and encouragement. Evarts came to Lyon early that year to present a paper in the sleep science symposium that Jouvet had organized. On that occasion, Evarts advanced the first functional hypothesis of sleep that took brain cell physiology into account. Because Evarts was one of my scientific heroes, I dedicated my 1989 book, *Sleep*, to him and my other NIH mentor, Fred Snyder.

In the early 1960s, budding brain scientists like me had a relatively easy time getting financial support. I had a fellowship to work with Jouvet, which continued as I finished my psychiatry residency and worked in Harvard's psychiatry department half-time over two years (1965–1967). Now it is practically impossible for young scientists to forge such a career. The fellowships are gone, and it's much harder to get grants.

This is not to say I was not under pressure. I was. My scientific mentors were all ingenious and productive. My academic boss made it clear that "publish or perish" was the law of the land. The fact that I was able to equip a lab and obtain grant support fairly easily only meant that I was bound to come up with something original. I myself hoped that something would be paradigm-shaking.

The NIMH was generous to young sleep scientists because external advisers had earmarked our fledgling field as particularly promising, an evaluation that now seems as prescient as it was fortunate. But the vaunted peer review system tended to reward slow, steady progress and to be leery of great leaps forward like the one I wanted to make.

So for the first 18 years (1968–1986), I did neurophysiology with an almost maniacal intensity. Bob McCarley and I often recorded neurones from early morning until 11:00 p.m., and even then we did not quit if things were going right. One of us might go home to sleep 4 hours and then return to the lab to spell the other at the controls of a microelectrode recording of a single cell in the pontine brain stem. These recordings sometimes lasted as long as 18 hours! Throughout this early work period, Ed Evarts was my inspiration because, as a psychiatrist himself, he realized the full promise of sleep research. Ed said, "If we can't figure out what's going on in sleep, how can we hope to solve the problems of major mental illness?"

In the face of my open admiration for Evarts, he always remained an aloof and distant figure whom I feared as much as revered. This is transparently obvious from my attribution to him of a critical dialogue regarding my mental and physical health. When he asks in the dream (which he would *never* do in real life) if I have had a mood swing, I am using him to ask a question of myself. 1986 was a critical year in the breakup of my first marriage because my wife, Joan, discovered the extent and details of my extramarital affairs by reading my journals. So this dream is a revelation of my concern about myself. If I did not have a mood swing (and I did not), part of me believed I should have or could have!

As for the herpes sore, my dreaming mind uses it to explain Ed's distance. I could never figure out why he was so aloof. As is usual in many such cases, I thought it must have something to do with me. "Don't come near me unless you're having it treated" is a strange comment anyway. Would I be any less infectious if I were using Zovirax (which I learned about only after 1994), and why would I be infectious to him, anyway? Clearly, I wanted to be closer to him than he did to me! A lot of other people had the same reaction to Evarts.

Scientific father figures are few and far between. A young investigator is lucky to find one. But I was lucky enough to find several. Two of them, Elwood Henneman and Vernon Mountcastle, were physiologists, not psychiatrists, but they were fascinated with sleep. They eschewed sleep as a scientific problem because it was risky and because it was not easily addressed within the Sherringtonian reflex paradigm that guided their work. But they were willing, even eager, to see me walk the plank and cheered when I stayed out of the water.

My other heroes in those early days were David Hubel and Torsten Wiesel, who used the same single-cell recording technique to dope out the code that the brain uses in early stages of visual processing. They went on to win a Nobel prize. Wiesel was a Swedish-trained psychiatrist. He didn't understand the state concept that I was drawn to. Hubel was a neurologist who had set out to study sleep but found it too messy to pursue. Hubel as much as told me I was on an impossible mission. The NIMH soon came to underestimate the value of investigating dreams. Because of the failure of the one-to-one brain-to-mind model that guided dream research from about 1955 until 1975, the funding sources for studying dreams dried up. Committees reviewing grants said, "Keep doing the neurophysiology, but stay away from dreams. Dream research is a scientific graveyard." After our own discoveries leading to the 1975 model of reciprocal interaction and the 1977 activation-synthesis dream theory, this was discouraging.

I hope time will prove me correct in ignoring these pessimistic prognostications. Dreaming, seen as a brain-based state of consciousness statistically correlated with REM sleep, has to be studied within the paradigm of bidirectional mapping. We have shown that dreaming *can* be studied scientifically and that its study is a signal of a new era in brain-mind science.

There is probably something significant in the fact that so many of my supporters were psychiatrists-turned-physiologists. Psychiatrically trained physiologists have their clinical experience within them as they scan the brain for relevant data. They look for things that other scientists deny and see things that others may even ignore.

The person in my life who best exemplified this psychiatrist-as-physiologist principle was Eric Kandel, who decided early in his career to focus on learning—as basic to memory—and to analyze this process in a radically simple system, the sea slug, Aplysia. Eric and I were first-year residents at the Massachusetts Mental Health Center in 1960. I remember Eric leading a hardy handful of us through the Ionic Hypothesis of Hodgkin and Huxley while our psychoanalytic mentors were advising us to listen to our patients and stay out of the library! Eric's lectures were better than the Research Seminars that Milton Greenblatt devoted to the personal histories of his scientific colleagues as a way of discerning psychodynamic aspects of their motivation.

We felt like we were at a new dawn of enlightenment and that we would lead the psychiatric world out of its superstitious and religious fixation on psychoanalysis. In that respect, Eric has vindicated us all by winning the Nobel Prize for Science and Medicine. But Eric is still working within the reflex paradigm and therefore is hostage to the closed-loop view of the brain-mind that hobbled Freud so badly.

All these associations constitute a kind of psychoanalysis of this dream. I find these interpretations interesting, plausible, and useful, even if I can't prove their validity. However, an important point is that even these interpretations aren't symbolic decodings, nor do they have anything to do with my unconscious. The sensitive reader will note that I stop short of assuming that I want to kiss Evarts to explain the herpes dialogue! I must say that although I don't mind kissing men, the thought of kissing Evarts is anathema to my conscious mind.

But as easy as it is to understand Evarts' presence and behavior in this dream, it is not at all easy to see how Mickey Mantle fits in. He shares two historical features with Evarts: hero status and Washington, D.C. Mantle's hero status goes way back in my psyche. Back beyond Washington, D.C., where at the same time I was beginning to do sleep research at the NIH, Mantle and Roger Maris were chasing Babe Ruth's record of 60 home-run swats in a single season. I have been a Yankees fan since 1941, when I was 8 and the Lou Gehrig legend still flourished. Gehrig was called the Iron Horse because he never missed a game. I admired Gehrig and wanted to be a baseball star long before I decided to go to medical school. At age 8, I was fairly sure that when I grew up I would play first base for the Yankees.

In this respect, living in Hartford, Connecticut was a problem for me, because although it was equidistant between New York and Boston, most New Englanders assumed that I should be a Red Sox fan. My resistance was unyielding. The Yankees were exciting. And they won. The Red Sox weren't exciting. And they lost. These things mattered to me as a young man.

Mickey Mantle insinuated himself deeply in my mind during the summer of 1962, when I lived in Washington and worked on sleep at the NIMH with Evarts. As I have said, our first child, Ian, was born at Walter Reed Hospital on February 1, 1962, while I was doing an all-night sleep recording in Bethesda. I thought that Ian, who was 24 at the time of the dream, would have enjoyed meeting Mickey Mantle even though he was an avid Red Sox fan (so much for family tradition).

I vividly remember getting up every morning in that long, hot D.C. summer and rushing to the door of our Tunlaw Road apartment to get my copy of the *Washington Post* and learn about yesterday's home-run production. In September, as the race got down to the wire, I even made a trip to New York and saw Roger Maris hit two out of Yankee Stadium in a single game with the Detroit Tigers. They were both towering blasts to the upper deck in right field. Like many people, I wanted Mantle to break the record and was sorry that he didn't. But because Maris was a Yankee too, that made it okay.

Mantle's presence at a grant committee meeting can best be explained by the habit of long-term memory to make associations on the basis of temporal and spatial proximity. In this case, Ed Evarts and Mickey Mantle are associated because of their temporal and spatial coincidence in my long-term memory. Time and space are important.

Many neurobiologists who aspire to a bottom-up theory of memory use that term interchangeably with learning. Learning is undoubtedly related to memory but should not be confused with it. Learning is experience-based plasticity. Learning obeys the reflexive laws of Pavlovian and Operant Conditioning.

Pavlov showed that the temporal pairing of a neutral stimulus, like a bell, with an intrinsically rewarding one, like meat, could give the bell alone the power to induce salivation. Skinner showed that reinforcing any behavior with a reward increases that behavior's tendency to occur. Great progress has been made in understanding the neurophysiological basis of learning. Kandel's cellular and molecular analysis of Pavlovian conditioning in the snail is one recent and spectacular example. Pavlovian conditioning of the snail's reflex has been shown to alter the excitability of the nerve cells mediating the reflex. Furthermore, Kandel has shown that serotonin neurones must be excited for learning to occur. This is notable since in REM sleep, when memory is so radically altered, serotonin becomes unavailable.

Memory is the conscious recollection of past experience. Certainly our remembered experiences involve learning, and certainly our conscious recollection of learned experience may obey some of the laws of conditioning. But learning, especially in a snail, cannot be equated with human memory. As I recollect my experience with Evarts, I can see him and imagine him in a wide variety of settings. These images and the feelings associated with them are the mental elements of my memory of him. I do not believe snails do this.

I talked with Ed, mostly about science, in his office in Building 10 at NIMH. I remember witnessing his pain at the flirtatious behavior of his wife at a sleep research meeting at a motel in Alexandria, Virginia. And I remember bumping into him at many of the early meetings of the Society for Neuroscience. By then, he had separated from his wife, was living alone in a sparsely furnished flat, and was learning to speak and write Japanese. He swam a lot. And he stopped eating lunch. He often went with me to a restaurant, and we talked as I ate.

These are my conscious memories of Ed. Cognitive neuroscientists would say they were acquired in waking when my hippocampus and cortex were working in unison. The memories are now distributed in my cerebral cortex. So why were these memories unavailable to me when I dreamed about Ed in 1986? He was by then dead, and my dreaming brain didn't know that. When I am awake, I have no trouble summoning these memories. The trouble I have is stopping them. Once the gates are open, my mind is flooded with Evarts memories.

My colleagues, including Bob Stickgold and Gyorgi Buszaki, tell me that in REM sleep I can access only fragments of my Evarts memories because my hippocampus and cortex are not working together. In REM sleep, I have an "internal" amnesia in which I am cut off from my own abundant memory of Evarts stories. Instead, Ed appears as a very limited and distorted dream actor inviting symbolic interpretation.

But just as we have been inclined to question other received ideas about dreaming, let's reject this one too and work on the meaning of Ed's appearance in my dream. Does it help to suggest that, as disoriented as my Mickey Mantle dream is, it does have coherence if remote memories are stored categorically? In Washington, during the summer of 1962, several important things were happening. My son, Ian, was developing slowly. I was beginning to learn basic sleep research from Evarts, and Mickey Mantle was chasing Babe Ruth's home-run record. Now, in 1986, I have a 24-year-old son who loves baseball. I have a satisfying scientific career but still harbor a persistent need for heroes or at least champions of my cause. Even the great Mantle can be turned down. Disappointed. And rejected.

Notice that my definition of memory as the conscious recollection of past experience has several important implications. One is that memory depends on narrative, and narrative depends on language. What the cognitive neuroscientists call episodic memory can only be deduced from what subjects tell us they remember. This means that, by definition, no creature that cannot give an account of its internal experience verbally can be said to have memory. Here I will be challenged by those cognitive neuroscientists who refer to procedural memory as the store of learned behavior that may or may not be brought to consciousness. To them I say, simply use the term *procedural learning* instead of *procedural memory*. In so doing, we lose nothing but the illusion that we are studying memory when we are really studying learning.

What is undoubtedly true is that both learning and memory are, like consciousness, exquisitely state-dependent. When I am awake, I can learn. When I sleep, I cannot learn, and I cannot recollect as well as I can using my waking memory. The brain changes that underlie these striking differences are likely to help us understand how they mediated.

In my dreaming brain's need to establish some semblance of orientational order, it is science, not baseball, that calls the shots. That's certainly true of my life too. I embraced science and gave up baseball. But baseball and the New York Yankees continue to interest me, and that might help explain why Mickey Mantle appeared in my dream even 24 years after his temporal and spatial association with Evarts. And it might explain why Evarts "papered over" (not the normal term for a grant rejection) his grant application (his bid for the homer record?). Does my attribution to Mantle of a downcast look express my sympathy for his disappointment in 1962? And is my reassuring comment to him about my review of his grant my way of saying that I'm still a fan of his anyway?

Here I am again bordering on the psychoanalytic. Mantle's grant application is his displacement of his quest for the homer record. And my quest for grants is my displacement of a childhood wish for fame as a Yankee to an adult wish for fame as a scientist. Such speculative interpretations can be taken with a grain of salt without weakening either the formalist approach or its assumption of transparency in the construction of dream meaning.

One of the most interesting aspects of Mantle's presence in my dream is his superstar quality and my ambivalence about inviting him to stay at my house for supper. The only person I know who has helicopters waiting to whisk him away from scientific functions is Gerald Edelman, the Nobel laureate immunologist-turned-neuroscientist who has written extensively on the brain basis of consciousness. Obviously, Mantle was eager to go and/or I was eager to get rid of him. Despite the attraction of presenting him to my son Ian, I used good judgment in deciding not to invite him home.

I am still unsure about how to handle authority figures, whether they be scientific or popular, as I am fearful of criticism and/or rejection. If long-term memory is so uncannily accurate in associating Evarts and Mantle in my 1962 file record, why is short-term memory so poor in reality testing my 1986 dream plot? Why didn't I recognize, at once, that Evarts wasn't even alive anymore (much less fraternizing with Mickey Mantle at a grant committee meeting?!). I knew, as soon as I woke up, that this was the case. I could say that it was inconvenient for me to have Evarts dead when my career was in crisis, as it was that year, owing to my first rejection of a grant renewal by the NIMH! I first received that grant in 1968, and it had been renewed every three years until it was "papered over" in 1986. That's certainly emotionally salient!

When I use my waking mind, with its very different form of memory capacity, to think about my dreaming mind, I can recognize some important spatial and temporal associations that are used to codify my own experience in my brain. As much as revealing my personal predilections, this sort of recognition suggests more general—and possibly universal—rules of how the mind works. The revelation of emotional salience vindicates Freud's emphasis on feelings at the same time it turns his dream model on its head.

Chapter 7

Bicentennial Wine Tasting

The Brain Gets High on Its Own Juices

I am attending a reception hosted by the French consul or ambassador in Washington to celebrate the 200th anniversary of American Independence. It is a huge gathering, which seems to take place in the south wing of the Capitol (the House of Representatives). Many people are milling about, waiting for ceremonies to begin.

Suddenly, to my left, my old friend, Skip Schreiber (from Wesleyan days), is hoisted to the shoulders of this throng as if he were to give a speech in his capacity of Maître des Chais—or Chef des Caves. Surprised, I call out to him, in a high-pitched squeal of a voice, "Are you the outgoing or incoming chef?"

Skip, who is as fair, blond, and youthful as he was at age 21 in 1954 when I last saw him, smiles mischievously and points, with his thumb, over his left shoulder. This gesture clearly indicates that he is outgoing and that his replacement is coming up from behind, left.

The new chef is a huge, smooth-skinned, handsome stranger. He promptly announces that, to celebrate the occasion, he is offering a wine made in 1779. I am thrilled at the prospect of tasting such an old vintage, and I quickly maneuver to be sure of a favorable position when the pouring starts. I can see, in my mind—but clearly—a bottle 6 feet high and 2 feet across—as large as a person—and envisage the label to be a scarcely legible handwritten one. Perhaps it was laid down by George Washington himself.

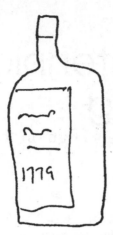

On a table, to my right, is a perfectly ordinary sized bottle—green with a bright red lead cap. I say to myself, "That will never be enough for this crowd."

A t 5:30 a.m., on a weekday while sleeping at home, I awoke and had detailed recall of this dream, which is notable for the way it integrates disparate elements and for its extreme bizarreness. The emotional salience of this dream is my love of larger-than-life celebrations.

The following antecedent events seem relevant to the plot construction. Two weeks previously, I had visited Washington, the Capitol, and the House chambers with my son, Ian. On Friday evening, we had dinner at L'Escargot, a French restaurant, with Alice Denney and members of her avant-garde art delegation. The waiters—in reality, not in my dream—wore roller skates! I bought two large bottles of white wine for our guests. It was plenty.

On the Saturday before the dream, we had dinner with our friends and professional colleagues Stanley and Judy Rapaport. I bought two bottles of Gigondas (at Plain Old Pierson's on Wisconsin Avenue), which they did not serve. At Pierson's I also saw a huge bottle, a double jeroboam of Bordeaux that was priced at only $99 and which tempted me.

Several days before the dream, I had been in New York for a television appearance on a morning show. At the studio I met R.S., with whom I had breakfast. She had a farm and wanted to live in the South of France. My fantasies and memories of France were stirred by this encounter.

Later, after a meeting with my book editor, I went to MOMA to see Paul Klee, but I gravitated to Odilon Redon's series of etchings called "In the Dream-Germination" and Henri Lartigue's photographs, which pulled together many of my favorite themes: French people, bed, women, cafés, wine, poetry, love. I cried at Lartigue's evocative memories. I yearned to be with R.S. and share this sensibility in an opening love.

R.S. lived in Washington. I thought of seeing her there, but she was cool and didn't come to lunch. So I went to the Brasserie restaurant alone, having missed contact with Lili Armstrong, another old friend, and ordered oysters (with a glass of Chardonnay) and duck paté (with a glass of Burgundy).

The evening before the dream, when going from our Brookline kitchen to the pantry, I saw the double magnum of Pichon Longueville (1980) that my French friend Jean Didier Vincent had bought me, and I thought, "How nice it would be to take this to Kate Kimmich's 50th birthday party," which was a week or so away. My wife and I had been discussing how to get there. I wanted to make a splash at the "elegant costume" party.

Vivid and richly detailed hallucinosis produces the complex and highly unlikely scenario of a French celebration of American independence in the Capitol Building in Washington. Fragments of all these elements, including the wine, are reasonably attributed to recent experience. But they are blended in what is nothing less than a creative synthesis that cannot be explained by mere associativeness. This dream scene is greater than the sum of its parts, as if it were in the service of the hyperbolic celebratory emotion that energizes it. So vivid and convincing were my perceptions and feelings that I never once questioned their reality. My false belief in the veracity of this dream convinced me I was awake. Even when I realized that the wine might not be plentiful enough for this crowd, I did not suspect I was dreaming.

We are so deeply involved with Freud's dream theory that we call all these experimental dream seeds "day residues." But Freud's idea—that each dream pairs same-day experience with an unconscious wish—is wrong in both its assumptions. The memory seeds are more likely planted many days before the dream, and the wishes they sometimes pair up with are not unconscious. They may also pair up with fears, not wishes. Michel Jouvet, a world traveler and dream journalist, says that new local events do not appear in his dreams for about a week following his arrival in a new place. The Canadian dream scientist Tore Nielsen says that same-day events can be included but that the peak of memory incorporation in dreams occurs six days after the events that seed the dream.

The orientational instability of dreams is strikingly illustrated in my dream. My rich Bicentennial experience—which never included such scenes as these—is transplanted to Washington (place inconstancy), where Skip Schreiber (person inconstancy), of all people, turns up. He has an unlikely role. Also, times are melded (events from 1955, 1976, and 1989 occur simultaneously).

My memory is clearly deficient. If it were working properly, I would immediately recognize this impossible fusion of times, places, and persons. In waking, I would say to myself, Where exactly am I? What day is it? And how could Skip Schreiber be here?—especially since he looks like he did in 1954. My emotionally salient associations are as strong as my memory is deficient, and we think we know why: As dorsolateral prefrontal cortex declines in influence, limbic lobe activation increases and comes to dominate cognition.

The distinction between learning, which does not require consciousness, and memory, which does, tends to suggest that language or the capacity to form propositional thought is a watershed in brain evolution. It is probably limited to the human mind, even if higher primates have some capacity to decode language symbols. Our capacity to frame experience in narrative form is greater than mere reportage. It is intrinsically mythical in that we cannot tell stories about ourselves that are completely accurate but, instead, elaborate our memories according to our own self-image.

Even myth-making falls short of adequately describing what consciousness can accomplish. Scenario construction is apparently innate in us. Our neurocognitive development between the age of language acquisition (3–5) and the age of adult fiction writing (15–30) shapes this remarkable capacity, which dreaming shows to be as universal as it is admirable. Dream consciousness takes pieces of experience and melds them seamlessly into totally enveloping, totally convincing, internally consistent scenarios of which we are the scriptwriter, the director, and the film projectionist.

Our proto-artistic capacity to create meaningful scenarios about ourselves is deeply natural and extremely useful. Orientation in time, place, and person is only a small part of what the brain must compute to promote effective waking behavior. We also need to be driven to specific actions by a very long-range view of ourselves that guides our behavior and helps us make critical life decisions.

At the time of my Bicentennial Wine Tasting dream, I was stretched to the limit of my investigative powers in the study of neuroscience. Cracks were appearing in my personal world. Important people were about to abandon me. But my consciousness was fueled and shaped by my story of myself that expressed itself, in scenario form, in my dream. In contrast to the Freudian explanation of the mind in terms of the perverse habit of disguise-censorship, we are now considering dreams as functioning to reveal, not conceal, personal truth. The mode of representation in dream consciousness is different from that of waking, but the meaning of that representation is at least as clear as it is when we are awake.

In order to refurbish the brain-mind via the forging of weak but emotionally salient links, the brain abandons its anchor in time, place, and person orientation, as well as its compass in executive function. These links guide our actions over the long term.

A word about confabulation: This word is used by psychiatrists and neurologists to describe the tendency of patients with recent memory loss to make up stories about themselves that are patently untrue. A person who suffers from Korsakov's psychosis might not know today's date or be able to say where he is, but he will fill in these blanks by making up stories. These are not conscious lies any more than the stories we make up about ourselves in dreams are conscious lies. These dream stories are, by definition, confabulations.

The capacity for psychosis is universal, and this capacity is revealed in REM sleep dreams. My point is akin to Freud's idea about the universality of neurosis, an idea he derived in part from his dream theory. The similarity of my idea and Freud's stands in stark contrast to our different interpretations of content. For me, dreams are revelatory and worthy of clinical interest because they illustrate, generically, how our minds work and how they express our individual psychologies in a generically constant way.

In my Bicentennial dream, there is plenty of evidence of defective thinking. When I cannot explain Skip Schreiber's role, I summon a stand-in, which convinces my dreaming brain that it is the chief wine steward of the entire country who is in charge of this national event. As if to prove it, I see "in my mind, but clearly" a wine bottle 6 feet high and 2 feet across that was laid down in 1779, "perhaps by George Washington himself!" Skip Schreiber is obviously filed in the celebration module of my memory. His appearance thus makes emotional sense even if it is cognitively impossible.

So we can add the principle that feeling is believing to the principle that seeing is believing. In fact, we can even postulate that our beliefs are a product not only of our percepts but also of our feelings. If I feel good about the existence of Santa Claus (or God, or astrological ordination) and I see what I take to be evidence of his existence (Christmas presents, the universe, and character), I can believe in those agencies with a conviction that is as powerful as my scientific convictions.

Most of our behavior is religious. Whether we are true believers or atheists hardly matters. There is simply not enough knowledge to base our world view entirely upon science. We plunge into the uncharted sea of experience and believe enough in ourselves to master unforeseen disasters. Belief is necessary in attempting anything out of the ordinary. And the sad truth is that organized religion discourages behaviors that are out of the ordinary so that most people who are true believers do not attempt transcendent, paradigm-shattering assaults on received ideas.

In the dreams of our most rational, most atheistic fellow humans, belief rules. This is because, absent the corrective structures of external space-time and the internal guidance of self-reflection, our dream perceptions and feelings convince us, falsely, that we are awake. This tendency to believe that what one sees and feels is real can be better held in check when we are wide awake, but the tendency is always there.

In other words, even atheists cannot help tending to believe in something. They believe in an illusory construct such as "self," the sensual external reality they perceive, or scientific discovery, which, after all, offers only extremely limited information about life processes. "Believe we must" is a categorical imperative that comes with our limited cerebral territory.

Some consider reflection on or interpretation of one's own dreams healthy narcissism. But narcissism is not always healthy, and "healthy narcissism" does not in any case capture certain wellsprings of energy and enthusiasm as well as the term *hypomania*. *Hypo* means just below the level of mania, and *mania* denotes psychotic delusions of grandeur and a disengagement with social constraint.

My brain is capable of cranking up my sense of pleasure and lowering my threshold for reward. My Bicentennial Wine Tasting dream demonstrates this aspect of myself. How does it happen? Why are some dreams filled with a sense of elation, joy, and happiness to the point of grandiosity? It cannot be due to the release of norepinephrine or serotonin, because their release is blocked in REM. More research is needed on the corelease in REM of acetylcholine, which we know to be robust, and dopamine, the molecule that brain science has earmarked for pleasure and elation.

Mark Solms, the psychoanalytically inclined neuropsychologist, likes the dopamine hypothesis because it helps him keep the Freudian notion of wish fulfillment alive. I like it because so many of my dreams have a sybaritic quality. But, as yet, the evidence for this idea is weak and the evidence against it is strong.

In 1987, it was already over ten years since the U.S. Bicentennial was celebrated. The Bicentennial was emotionally salient to me for many reasons. In 1973 my first wife, Joan, went to work for the Boston Bicentennial Celebration. Our third child, Julia, was born in December of 1972, and we needed help to take care of her and hold our household together. At the invitation of Jean Didier Vincent, my scientific colleague in Bordeaux, I went to France and began a long series of artistic enterprises. I also recruited Chantal Rode, a young French woman who had worked for the Vincents, to come to Boston and be our housekeeper. She stayed with us in that capacity for 23 years!

The point is to show that the Bicentennial, France, and wine are all tightly linked in my mind in an emotionally salient way. At that time, we knew the Boston-based French Consul, Roger Establie, and his wife, Suzanne, but we were never connected with his peer in Washington. The main setting of the dream in the U.S. Capitol must be ascribed to other associations. The principal site of the Bicentennial Celebration was Washington, and, as noted, I had recently visited that city and celebrated reunions with my medical school friends (the Rapaports) and another master of celebration (Alice Denney), whose parties for art world camp followers were legendary. The appearance of the South Wing of the Capitol reflects my recent visit to that building with Ian.

I first met Alice Denney at a party in the early fall of 1961 when I had just moved to Washington, D.C. to do my NIH stint. We enjoyed dancing together and thus began an exciting friendship that lasted forty years. Alice was then working with Adeline Breeskin, founding director of the Washington Museum of Modern Art. In this role, she knew many of the bright young lights of American Art, which was then on the cusp of abstract expressionism and Pop. On Sunday mornings, after openings, happenings, and other early '60s goings-on, Alice entertained artists and artistic camp followers like me at her house in northwest Washington, which was walking distance from our Tunlaw Road apartment.

It is important to note these facts because they contributed so strongly to the Washington, D.C. that I knew when I was first introduced to sleep and dream research. In a word, Alice Denney was a strong and welcome antidote to the stuffy, bureaucratic Washington of government agency fame. For me, she was an ally of our young president, John F. Kennedy, who was then in the midst of his famous 1,000 days of rhetoric and cage rattling.

Through Alice, I met Claus Oldenberg (of soft hamburger fame), Larry Poons (whose painted dots danced), and Robert Indian (whose graphic 5 is still etched in my mind). A combination of art and science thus seemed possible, and this was especially true of the autopsychoanalytic work of Jim Dine. Dine's co-option of the psychoanalytic paradigm and his mockery of it in his famous happenings were signs of decadence that we did not appreciate at the time.

Through Alice, I also met the great colorist painters Jules Olikski and Robert Natkin. Natkin responded to my instant appreciation of his work by selling me one of his Apollo series paintings for practically nothing. Not only that, but he brought the painting to Washington, D.C. by rolling it up and carrying it on the shuttle with him. We then drove together to Hecht's hardware store on Wisconsin Avenue to buy wood for a stretcher frame and stapled the canvas to it on the floor of my Tunlaw Road apartment. This painting has graced my life ever since.

The Apollonian myth is certainly present in my Bicentennial Wine Tasting dream. But so is the Dionysian side of me. I want to bask in the sun. That's why I go to Sicily. But I also want to work in the shade. That's why I stay inside and write. People like Alice Denney and Bob Natkin are not portrayed in this dream. But they are there, behind the scenes, coaching me on my dream scenario.

We can only speculate about the brain basis of REM sleep dream grandiosity and wake-state hypomania. We know that the emotion centers in the limbic lobe are activated in REM. And we know that those centers include the amygdala, which mediates fear, but also the deep frontal areas adjacent to the hypothalamus, which may mediate rage (which is common in dreams), elation (which is very strong in this one), and sex (more on this later). One reason that people are drawn to amphetamine and cocaine abuse is because those drugs make them high. They have this effect via their ability to simulate or stimulate dopamine release. Since dopamine release is not impaired in REM, it could play a role in tipping the brain's chemical balance in the direction of elation and hypomania, especially since serotonin and norepinephrine release is shut off in REM.

As in my Mickey Mantle dream, fused to this dream is a principal character, Skip Schreiber, my fraternity brother at Wesleyan, where I last saw him in 1954! How can his appearance be explained? Skip Schreiber is associated in my memory with formal elegance, dining, and drinking. He always dressed for dinner and played bridge brilliantly as he drank martinis before going to the dining room of our fraternity house. Skip was a bit of a peacock, as was the man who decided to become Maître des Chais in my dream. Why does the role of Maître des Chais change? Because Skip Schreiber is miscast? Perhaps. Certainly, his unidentified replacement is more compatible with this function. It is the new man who offers us the mother of all wines, the 1779 George Washington!

Although I had been introduced to fine French wines while an exchange student in England in 1951 and 1952, we didn't drink much wine at Wesleyan. The wine habit—and hobby—began in earnest when I went to medical school in Boston. It became intense ten years later in Lyon, where I worked with Michel Jouvet. By then I was well attuned—if not addicted—to the wines of Beaujolais, Burgundy, and the Rhône Valley. Later, in Bordeaux between 1973 and 1984, I learned a lot about the great reds of Médoc, the whites of the Graves, and the delicious desserts of the Sauternais. So, although I can't associate wine directly with Skip Schreiber, I can see a connection via the exhibitionism, the love of pomp, and the commitment to ceremony that we both shared.

As mentioned in my discussion of my Mickey Mantle dream, alcohol is, in fact, a powerful REM suppressant. In Lyon, where we often drank wine at lunch as well as at dinner, we joked about the Beaujolais effect. The Beaujolais effect was caused by too much wine of any type and was characterized by poor sleep early in the night, when the fusil oil and aldehyde breakdown products of alcohol adversely affected the brain. Later in the night, when the toxins were gone, the sleep-deprived brain went into overdrive, and the intense REM rebounds were accompanied by wildly bizarre dreams, like this one, which was not caused by wine but is about it.

When alcohol consumption becomes even more excessive, it can lead to even more severe disruptions of sleep. Then the withdrawal-related hangovers can only be treated by more alcohol, as in the famous "hair of the dog" remedy. The sleep suppression can ultimately become so grave as to lead to chronic REM deprivation and, as previously mentioned, true delirium—in this case, delirium tremens (DT)—which represents REM rebound that is so intense that it breaks through into waking.

I remember treating as many as six DT'ers simultaneously on ward P5 when I was an intern at Bellevue Psychiatric Hospital in the summer of 1959. To keep these people alive, we used ice packs to reduce body temperatures that threatened to rise above 106 degrees Fahrenheit. The patients ranted and raved, saw bugs crawling on the walls, and thrashed so violently in their beds that we were obliged to use restraints to keep them from hurting themselves.

Now that we know from extreme deprivation studies in animals that REM sleep is in the service of temperature regulation, we can better understand the dramatic temperature instability of our DT's patients. As soon as we managed to prevent the fever from cooking the brain (which occurs at about 108 degrees Fahrenheit), we needed to quickly remove the ice packs to prevent body temperature from falling through the floor into dangerous hypothermia. It is probably significant that REM sleep is the only brain state of mammals in which active temperature control is abandoned! Temperature is behaviorally regulated in sleep. We simply do not go into REM when the ambient temperature is dangerously hot or cold!

Grandiosity is clearly another theme of this Bicentennial dream. There is never enough wine! I need two bottles of wine, not just one. I am hoping to drink the two bottles of Gigondas, not just give them away to the Rapaports; I am tempted to buy a double jeroboam at Pierson's; and I already have a double magnum of Pichon Longueville—from Jean Didier Vincent—in my house. But it's never enough! My dream proves it. I see a person-sized bottle of incredible age—vintage 1779, with a handwritten label possibly written by George Washington himself! But I am worried that even this will be insufficient!

My sense of vinous inadequacy and my need for more, more, more are reflected in my hyperactivity during those days. I was in

Washington, D.C. with Ian one week, in New York City for a television appearance the next week, and about to drive to Norwalk for a 50th birthday party that weekend. My dream life during this period was as hypnomanic as my waking life.

Perhaps if I had analyzed the wine dream then, I could have caught myself. But I didn't want to catch myself. I was having too much fun. And I was managing—just barely—to keep my laboratory functioning, my family together, and my creativity on track. I say just barely because by then it was clear that my wife, Joan, was disaffected with me and that my coworker, Bob McCarley, needed to go his own way.

I am larger than life in my quest for recognition, celebration, and social interaction. Wine is the lubricant of many of the processes I foster. I love parties. I always have, and I always will. The combination of American know-how, independence, and can-do swagger with French sensuality, eating, and drinking has always been a major integrative factor in my life. My brain is obviously tilted in the direction of high energy, high pleasure, and high elation. Does this mean that my waking hypothalamus is overactive? Or that my waking brain is soaked with dopamine? These hypotheses are only indirectly supported by the findings of sleep and dream research. Fortunately, I am very sensitive to alcohol toxicity, so there is no danger of overdosing on that kind of juice.

Chapter 8

Dangerous Diving

The Surprising Importance of the Vestibular System

I am standing beside and behind a huge railway trestle—maybe 100 feet high—that is made of timbers and runs parallel to the ocean. High waves are breaking over the top of the trestle. Then I notice people diving into each wave as it breaks. They do gainers, pikes, and all manner of twisting, turning maneuvers. I think, "My God, how dangerous! Their timing must be perfect." To cushion their fall, they have to hit the wave before it rushes down the beach to the sea.

*The scene abruptly changes to a surf caster who is down
the beach to the left. He has landed a 2- or 3-foot trout. I
cry out, "Peter (Thompson)! Look at that!" The fish is
placed in a plastic tub and begins to swim in furious circles until it appears as a shimmering ring of movement
and almost disappears.*

*The fisherman puts his catch into a pond-like basin on the
beach, and it streaks away with lightning speed. "He'll
never catch it again," I lament.*

*Again the scene shifts, and I behold a host of fishermen,
tiny and distant. Some are on the beach at water's edge,
casting. Others are on the edge of a submarine canyon,
below the water's surface but plainly visible. They too are
casting.*

*I think, how wonderful. They have no masks, no diving
equipment.*

I was sleeping at home and woke up at 7:00 a.m. needing to urinate. This setting is typical of conditions in which I have rich,
vivid recall of dreams. However, it is not typical for me to sleep
fitfully, with many partial awakenings, so I might have been partially REM-deprived. This would explain my notation of this as an
"intense, brief, strikingly hallucinoid" dream.

It also might be significant that I was reading a biography of
Salvador Dali, the surrealist artist whose paintings have fascinated
me for years. I find Dali's imagery to be the most convincingly hallucinoid of all the surrealists, and the chapter I had been reading
emphasized his treatment of the sea. The ambiguity of figures on
beaches, the transparency of his painting surface, and the episode
of his London lecture, when he wore a diving suit, are all relevant
to understanding this dream's bizarreness. Like him or not, we
must admit that Dali celebrated the bizarreness of dreams in his
life as well as in his art.

Although I am safely planted on terra firma, my dream vision beholds a perilous and improbable scene. Railway trestles do sometimes run close to the sea, but they are not normally founded in sand. And because boys will be boys, they sometimes dive from the most hazardous heights into water of uncertain depth. But by diving into waves breaking on the beach, my dream divers do something much more dangerous. My dream railway trestle thus defies reality. Surf of the scale dreamed by me would soon bring down any real trestle, just as dives into rapidly receding waves would soon undo the divers, too. But the hallucinoid intensity of this graphic image gives rise to only token fear. "My God, how dangerous," I say. The faulty dream logic is used to explain this anomaly. "Their timing must be perfect" is a perfect example of what I call an ad hoc explanation. Ad hoc explanations are the often-weak cognitive responses we make to unlikely dream imagery.

Not only is this opening scene as vivid as a Dali seascape, but also it is animated in a way that a painting can only suggest via a multiplicity of images. My dream waves crash into the trestle, and the divers hurl themselves into the crest of the breaking waves. My brain must be motorically and visually active, and the combination results in what I call visuomotor hallucinations. Sometimes it is I who moves. Sometimes, as here, it is the dream characters. Not only do these divers defy death, but also they do so elegantly as gainers, pikes, and twists decorate their plunges.

In my discussion of modern neuroscience, I have stressed the important role of the vestibular nuclei of the brain stem in REM sleep eye movement generation. Understanding more about this well-studied system can help us better appreciate this visuomotor imagery of dreams, which is nowhere better illustrated than in my Dangerous Diving dream. Let's start with the Activation-Synthesis hypothesis and work our way back to basic neuroscience. In 1977, Robert McCarley and I suggested that all dream movement and especially the spectacular trajectories of stunt divers, circus acrobats, amateur flyers, and all manner of dream spinners was derived, in part, from the offline activation of the vestibular system.

In waking life, it is the job of the vestibular system to track every detail of the body's constantly changing position in space and to integrate that data with the position of the head and eyes. This is all done so effortlessly and so unconsciously that we take it for granted if we recognize it at all. But as anyone who has ever experienced motion sickness knows, the vestibular system is a very delicate and important piece of cerebral real estate. By riding on a boat in rough weather or by taking a spin on a circus joy ride, we can easily upset this system to the point of nausea and vomiting. And if you are ever so unfortunate as to have Meniere's labrythitis, which afflicted my mother-in-law, or suffer brain stem stroke, as I recently did, you will know that the system can be upset so that the outside world seems to spin even though it (and you) are perfectly still.

When your brain enters REM sleep, the neurons of your vestibular nuclei begin to fire in high-rate, rhythmic clusters that are associated with eye movements, just as they are in waking. This activation is in the circuit first described by Rafael Lorente de Nò in 1933 to explain the vestibulo-ocular reflex (VOR). This means that whenever the head moves, the eyes make a reflex adjustment in an effort to hold the visual field steady. But sleep activation of the vestibular system is not reflexive at all because no stimulus-related excitation of the circuit arises in the vestibular nerve, which carries impulses about head position from the middle ear during waking.

Thinking hard about the offline activation of the vestibular system in REM sleep is a good way to appreciate what goes on in the brain when we dream. It helps solve the problem of what constitutes the dream stimulus, and it leads to an immediate apprehension of the concept of fictive motion. Finally, it provides a solid scaffold for the development of the bottom-up dream theory I am advancing. By bottom-up, I mean the reasoning that proceeds from the bottom (brain physiology) up (to dream psychology). This is what Freud wanted to do but could not because of his limited knowledge.

If the brain is self-stimulating in sleep in a way that even remotely resembles its stimulus-driven activation in waking, the result is a convincing sense of movement of the self or of dream objects in the dream space. This is what I mean by fictive movement. It is the illusion of movement caused by the activation of brain structures associated with real movement in waking. We have already considered this idea in thinking about dream vision (which we ascribed to wake-like activation of the visual forebrain in REM) and dream emotion (which we ascribed to the activation in REM of the limbic system of the amygdale and related temporal lobe cortex).

So far, the only evidence of vestibular activation in human sleep is the PET-revealed selective activation of the pons. We would expect the better temporal resolution of fMRI to show selective activation of the vestibular system and cerebellum during the eye clusters of REM.

Why, then, are some dreams, like this one, so dynamically animated, and others, like my Mickey Mantle dream, so static? The obvious answer is that vestibular activation in sleep is highly variable. A more-specific hypothesis is that the intensity of perceived movement should be correlated with the intensity of recorded eye movement. This experiment can be done using existing technology. The Nightcap home-recording system has an eyelid movement sensor that is exquisitely sensitive to REM sleep eye movement.

Here we are revisiting the old psychophysiological paradigm, which posited a one-to-one cross-correlation of a mental state feature and its supposed physiological basis. But our hypothesis is much less demanding than the relationship of the direction of hallucinated gaze in the dream to the direction of the eye movement recorded in the electro-oculogram.

So far we have answered only mechanistic questions. What about function? Am I, like Greg Louganis, rehearsing my dive repertoire in my brain? I haven't done any of those dives in thirty years, but I haven't played baseball either, and I still know how! At least my brain knows all the moves, even if my creaky body doesn't follow my brain's instructions. Once again, we confront the attractive idea that the brain circuits so carefully tuned by early procedural learning practice are kept fresh by nightly reactivation. It's a good use to make of the downtime of sleep. We can use it to run the brain's programs for integrated behavior!

As one rapid eye movement cluster ends, another begins, and my dream vision shifts gaze from center to the left of the scene, where I see the surf caster landing a 2- to 3-foot trout. Without any explanation of his presence, my old friend Peter Thompson is the target of my exclamation of surprise: "Peter! Look at that!"

Surprise is a very common dream sensation. My colleagues have debated whether it should be considered an emotion. No wonder we are so often surprised in dreams, when so many improbable and unexpected things go on. But couldn't it be the other way around? That is, perhaps the dreaming brain is neurophysiologically primed to generate startle responses and thus needs to account for them at the level of conscious experience. I think this hypothesis, first advanced by Adrian Morrison at the University of Pennsylvania, needs to be taken more seriously.

Surprise is what we feel when we are startled. We jump when we process an unexpected stimulus. "Oh, you startled me" or "I almost jumped out my skin" is what we might say. This normal startle response is mediated by the same brain stem neuronal reflex circuits that become spontaneously active during REM sleep. This means that the brain-mind, in REM, is repetitively startled—that is, surprised—by its own automatic activation. As I lay in bed on the morning of May 19, 1987, I was surprised, again and again, because I was reacting subjectively in my dream to repeated startle reflexes arising in my brain stem. If my wife had been awake, she might have seen me twitch and might have heard a muffled cry when I called out to Peter in my dreams. And she would certainly have been able to perceive the clusters of REM behind my closed lids. At 7:11 a.m. in May, it has already been light enough to see the REMs for at least 1 1/2 hours.

My dream surf caster is successful in landing a 2- to 3-foot sea trout. When the trout is placed in the pail, it is even more animated than the divers. But this time the movement is circular instead of parabolic. This fish whirls so fast in the tub that it almost disappears "in a shimmering ring of movement." Sleep and dream researchers have speculated that circular movement is another product of brain stem vestibular activation. It is the vestibular part of the brain that normally tracks the position of the body in space, and it is this region that is at the heart of the REM generation process.

The damage to my own brain stem by stroke, which occurred in Monaco on February 1, 2001, was centered on the vestibular system, which is why I still have difficulty with balance. And that might be part of the reason I stopped dreaming for so long after I was taken ill. In my Dangerous Diving dream, the dream fish is still moving fast when it is put into the pond-like basin, and it streaks away—to my great dissatisfaction. Why it is sometimes I who moves and sometimes the observed fish is unknown. But what is sure is that in both cases the motor programs of my brain are being run, and it seems likely that they provide an important formative stimulus to animated dreams like this one.

In the last scene, I come close to Dali's surrealist vision when I see small fishermen standing underwater on a distant beach. The distance is improbable and the immersion is impossible, but I have no trouble accepting both illusions in my dream. I am faintly aware that this experience can't be veridical when I exclaim in wonder, "no masks, no diving equipment," but this does not cause me to question my perception. Here again, seeing is believing, as Dali himself so clearly recognized. If you paint something impossible that looks real, it *is* real, or even more than real. That is what we mean by the word surreal!

These dream themes are all related to my love of gastronomy and exhibitionism so clearly illustrated by the Bicentennial Wine dream. I loved going to Pied-a-Mer, Roger Prouty's house in Orleans. Roger, even more than Skip Schreiber, loved to dress up. He had a closet full of outfits that he also shared with his guests. The night that Peter Thompson and I made our sea bass sevichè was typical.

After a day of beach running, jeep riding, body surfing, and playing eccentric games of Roger's invention, we all enjoyed packing into the tiny sauna, cooking ourselves till well done, showering, and getting the group dinner ready. We drank martinis and other potent REM-suppressing cocktails, clowned outlandishly around the bar, and ate sumptuously, all the while dressed up as whomever—or whatever role—we wanted to play. After dinner, we staged tableaux that were every bit as imaginative as my dreams, but more daringly sexual and at the same time more subject to censoring. I was consciously uncomfortable with the invitation to exceed socially acceptable limits of decorum. This kind of conscience intervention never darkens my dream door!

Cape Cod and the islands, especially Martha's Vineyard, were more than a playground for me during my 20s and 30s. I had met my first wife, Joan Harlowe, on a blind date in Edgartown, Massachusetts on July 4, 1953, the year that REM sleep was discovered by Eugene Aserinsky in Chicago. Because we were both products of the Depression and puritanism, Joan and I were only recent ex-virgins when we married in 1956. Joan loved the beach and was willing to dress up with me so we could crash the Edgartown Yacht Club, drink stingers at $2.50 a glass, and dance to Lester Larin live until 1:00 a.m. Then we took our Cinderella coach back to the Harbor View Hotel and turned into lowly servants—she a waitress and I a bellboy.

Joan and I went along with the dream-like shenanigans at Roger's Pied-a-Mer because they were voyeuristic and mainly innocent. Body painting, along with Marquis de Sade tableaux, were about as daring as it got. The sexual revolution was about to break over all of us, and experimentation was just around the corner.

When I had the Dangerous Diving dream, my participation in the MacArthur Mind-Body Network was just around the corner, too. Going back to Edgartown with that group 35 years later, and staying—as a guest—in the Harbor View Hotel was a Proustian transformation for me. And catching a record-breaking fish—despite the nausea of seasickness—was a dream come true.

Interestingly, we never experience nausea or vertigo in our dreams when our vestibular circuits, but not our labyrinths, are activated. When we sleep, we do not experience a mismatch between the external world and the internal representation of it because we have only the internal representation to work with. Thus freed of reality's controlling hand, we run free as our dream dives, dream fish, and dream paintings unfold luxuriously.

I have always wanted to be a circus acrobat or a springboard and platform diver. When I was a boy of 8 or 10, my father helped me construct a wooden springboard in the field near our house in West Hartford, Connecticut. He also bought me a copy of *The Tumbler's Manual*, which had well-executed line drawings of stunts that could be performed on the ground or from the springboard. Talk about dangerous! My merry band of childhood helpers and I could easily have broken our necks. I would never let my own twin boys indulge in such antics. We had only a pile of hay to cushion our fall as we performed flying somersaults in the air! My mastery of the running front somersault was a deeply satisfying achievement that served my pride and exhibitionistic nature all through adolescence.

The beach (like the lobster in its dream) was too far away for the likes of poor old us. And the town swimming pool, with its diving board, was often declared off-limits because of the polio epidemic that was still a threat to youths in the early 1940s. So I made do with the makeshift springboard without ever thinking that it was far more likely that I would break my neck by tumbling in a field than get polio by swimming in a pool! We put trapezes in the trees, too. And I prided myself in hanging upside down by my toes, heels, and knees as we swung to and fro. I was in a Fellini movie. I loved the acrobatic aspects of life and even of science. I wasn't happy unless I was on the edge where ignorance and knowledge meet.

When I went to Loomis, there was no swimming pool, so I never got to dive competitively. But I did dive at Little Tuckahoe, the house of my girlfriend, Blannie Dew, in rural Connecticut near Torrington. Self-taught, I learned to do front and back gainers and pike position front and back somersaults. In all that risky experimentation, I hit my head on the board only once. Were all the motor programs that my dangerous dream divers executed stored in my brain? Were they waiting there for their revelation in a moment of fictive glory in a dream forty years later? Was I reinforcing this early motor learning in case I might need it again? Maybe. We know that recently acquired motor skills are consolidated or even improved in sleep.

Fishing was also a great draw to me, but I didn't land a big catch until some years after this dream occurred. And I never tried surfcasting. However, I did see a lot of it at Roger Prouty's house on Nauset Beach in Orleans. In fact, I often went there with my dream co-fisherman, Peter Thompson. On one occasion, we were given a good-sized striper by a successful surf caster. As a culinary experiment, we skinned the fish, marinated the fillets in lemon juice, and ate them raw as striped bass seviché.

What about the underwater fishermen—without masks or diving equipment? Breath-holding and underwater distance swimming were two of my other adolescent aquatic stunts. In my prime, I could do three successive 60-foot pool lengths underwater without surfacing. Like my flying front flips, this stunt set me apart from my peers. Harry Houdini inspired me. Fishing underwater—even surfcasting underwater—is not out of the question for a dreaming mind like mine!

As far as exhibitionism is concerned, Dali and Fellini are inspirational. I have always admired Dali's extraordinary talents as a painter and was interested to read about the attention he drew to himself by his provocative public behavior. There was something offensive, vulgar, and even disgusting about Dali's narcissistic showing off, but what a marvelous painter he was! And I could see the strong links between his peculiar paranoid-erotic philosophy and his compelling pictorial imagination.

The fact that I was reading Dali's biography when I had the Dangerous Diving dream seems significant in several ways. It awakened the acrobatic performer part of me. It helped me to create a magical fish. And it seeded the undersea fishermen in the final scene. So this dream can be viewed as a tribute to Dali even though he doesn't appear in it.

For the traditional Freudian psychoanalyst, these dangerous dives might be dream symbols of something else—of sexual activity, for example. And this might be what the "big trout" is all about. My censor is aroused and translates a sexual drive (the latent content) into disguised dream imagery (the manifest content). It has been argued that this kind of reasoning is a caricature of Freud's theory and/or that no one thinks that way anymore. And yet, the kernel of disguise-censorship lives on at the heart of questions such as "What does diving bring to mind?" and "What do you think the fish really means?" My answers to these questions might seem prosaic: Diving brings diving (and related stunts) to mind, and the fish brings fish (and related fishing scenes) to mind. It is certainly simple. Using the philosophical tool to gut complex explanations when simple ones will do, activation synthesis is to be preferred to disguise-censorship.

What would a Jungian say? That my anima is a soaring bird, or a death-defying spirit, and that my shadow is the fear of destruction, of being caught, trapped, and deprived of my transcendent freedom. So far, this is not so bad, but when Jung goes on to posit the collective unconscious (as in a universal wish to fly) or an archetypal upsurge (as in the emergence of shamanism), I part company with him. Freud rejected Jung because of his mysticism. While I much prefer Jung's dream theory to Freud's, his strongly antiscientific position is alarming. As a scientist, Freud is preferable.

We know that the brain stem's motor pattern generators are activated in REM sleep and that these motor generators activate the systems that govern our sense of position in space. Movement, even exotic movement, is the order of the night. That order includes flying and spinning. If *this* universal process is the "collective unconscious," so be it, but that is not what Carl Jung had in mind when he coined the term! Although it's superficially more compatible with Activation-Synthesis, Jungian dream theory is less attractive than the now-absurd doctrines of Freud, because Freud was at least tacitly aware of the importance of neurophysiology. Jung went to medical school and was an experimental associationist during his psychiatric residency with Manford Bleuler at the Burgholzi Clinic in Zurich. But Jung was a thorough-going mystic spiritualist. These traits may rightly endear him to artists, but they are incompatible with science.

In saying that my dreams amplify and clearly display important themes of my life, I am agreeing with contemporary dream content analysts such as William Domhoff, who minimize the differences between waking and dreaming consciousness to press for the continuity of individual traits across those state boundaries. My dreams don't prove Domhoff's hypothesis, but they are certainly consistent with it.

There are many reasons to hope that Domhoff is correct in his hypothesis that each of us dreams his own dream and that our dreams reflect our waking concerns. But we have good reason to be skeptical. We need to know, for example, if independent judges could sort, say, 100 dreams from ten dreamers into ten separate piles if they did not know which dreams came from which dreamers. We wonder, too, if even the dreamers themselves would recognize the reports as their own if their provenance were disguised and several years had passed since the dreams were collected.

I know my dreams are mine because I see the reports written in my hand in my journals, but even then, I sometimes marvel at the conclusion that I must have had this or that dream. Such wonderment arises because I have no conscious recollection whatsoever of having had that particular dream. Would I be able to identify my own ten dreams if the reports of them were mixed with ninety dream reports from nine other people? The specific names of people, places, and actions would need to be removed, and the writing style would need to be standardized to prevent the judges from grouping the reports according to literary style.

If these experiments failed—and they might—where would the Domhoff hypothesis be? Out the window with all the others. Yet it would remain psychologically useful to associate with dreams. But any dream—mine or anyone else's—could be as good a point of departure as any other. None of us wants to believe it, but it still might be true.

Another reason for being skeptical of Domhoff's claims is that he denies any specific formal differences between waking and dreaming consciousness. For Domhoff, dream vision is the same as mental imagery in waking. But the constant delusional belief that one is awake while actually dreaming contrasts with the very rarely asked question in waking: Am I dreaming? The bizarreness, especially the transmogrification of dreams, is almost qualitatively distinctive. Transmogrification is Martin Seligman's wonderful name for what we call changes in identity by dream characters or objects. A rope becomes a snake, a man becomes a woman, and one hotel room becomes another. The marked impairment of episodic memory limits dream access to facts that are readily available in waking. In the face of such glaring differences as the neuroimaging revolution provides, waking and dreaming seem quite distinct states.

Perhaps Domhoff does not recall his own dreams. Relying on the reports of others, especially students, is no substitute for the expert witness or trained observer approach we take, especially in designing pilot experiments, because many students are poor at reporting on any aspect of their mental life. With our pilot data in hand, we can create affirmative probes for our naïve subjects to use in writing reports that focus on one or another formal feature of dreams.

There is no necessarily exclusive difference between the conclusion that dreams reflect waking conscious and that dreaming is formally quite different from waking. After all, the biographical database is the same for every individual waker-dreamer. But to suggest that dreaming is as banal as waking is retrograde with respect to cognitive neuroscience. When my brain-mind is turned on strongly by early-morning, post-deprivation recovery REM sleep, it often synthesizes spectacular scenarios like those of dangerous diving. I then fulfill many of my lifelong yearnings: to dive, to be an acrobat, to defy death, to fish, to swim fast and move fast, and to be an artist—like Dali—who realizes the extremes of imagination. Only neuroscience can hope to explain these propensities.

Chapter 9

Split Whale
Fear and the Amygdala

I am with my son Ian, walking toward a decline leading to a lower field. I notice some vague humpbacked forms and say, "Oh, there are the bulls." It then occurs to me that we might not be wise to cross the fence and field just there, but simultaneously I feel reassured. "Oh, it's OK," I think. I am (we are?) carrying a very heavy and very peculiar fence post consisting of a normal cedar log and a plastic tubular bottom. At that instant (or later) I realize that the purpose of this bizarre object is to have strength above the ground (the log) and resistance to rotting below (the plastic tube).

The short incline has suddenly become a long, steep cliff falling precipitously to water far below. For a while we descend carefully from rock to rock, but the slope becomes increasingly perilous. I decide to jettison the post, and it falls endlessly before hitting the surface. I suddenly realize that it might injure someone on the beach below. However, instead of people, there appears to be a huge whale carcass, hanging vertically and split down the middle. Nearby, I notice a foxhole-like hut that has been made from chunks of whale blubber.

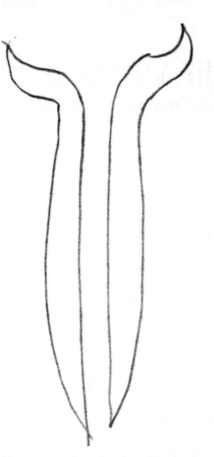

I realize that we must turn back, and I wonder if we will be able to ascend the steep slope. Looking down, I notice a wonderful slide just below, where penguins are playing gleefully. One large bird cavorts over the rocks, shaking water from its wings.

I slept at home with my wife Joan on the eve of a seminar on the Physiology of Dreams that I gave at St. Luke's Hospital in New York City one winter nearly two decades ago. I woke up at 5:40 a.m. with vivid recall of the dream transcribed previously. This early awakening could be related to my anticipation of the seminar, but the content is not in any obvious way connected. Because the last paragraph of the report is scrawled on the letter of invitation to the seminar, I assume that I wrote this account in Boston's Logan International Airport or on the plane going to New York.

The spatial orientation and emotional tone of each of the first three scenes is quickly established. I am with my son Ian on a hillside, as at our farm in Vermont, but this is not specified as such. This is typical of dream landscapes—they have the formal features of specific places in our lives but often differ in detail; it is as if "country landscape" were a generic mental category. This opens the associative net to bulls, poorly visualized but clearly menacing, and to synthetic fence posts, the better to contain the bulls! My dream invention of plastic-coated fence post is incongruous and creative. It *could* be done. Ten years later, and with no recollection of this dream, I decided to install a vinyl fence on the front edge of my Brookline garden. It will never rot.

Then fear gets in the way of my dream fence post project. The slope gets steeper and changes character to that of a seaside mountain slope like Big Sur in California. I therefore decide to jettison the heavy fence post but almost immediately regret this decision. The hallucinoid intensity of the cliff and the fear of falling from it have taken over my ability to reason or to use good judgment. This erosion of thought by hallucination is a typical feature of many dreams. As in mental illness, we can't think straight if we are seeing things.

As the dream proceeds, the emotional intensity increases. It is not just fear of falling, and the fear of hurting someone, but a more pervasive anticipation of violence that brings on the split whale and blubber foxhole images. A beached whale is uncommon, but I have seen them. But I have never seen a whale carcass perfectly split in the long axis. The blubber foxhole is a superbly condensed incongruity. In human wars, foxholes are dug on beaches for protection. It is protection that this whale (and all whales) need. Perhaps I, as the frightened dreamer, need it too. While the whale blubber could be used to pad the walls of a foxhole (and that could offer some protection), this is a most unusual architecture. My brain-mind is being activated in such a way as to demand integration of fear of violence with beached whales. And so it does.

With cognition going from bad to worse in my dream, I decide that the descent is too risky. My reasoning power returns at the same time the vertiginous feeling—and the threat of falling—declines. This is due to spontaneous abatement of my brain stem's vestibular activation process (resulting in vertigo decreasing), the decline in activation of the amygdala (resulting in fear decreasing), and the partial recovery of dorsolateral prefrontal cortex (resulting in an increasing capacity to think rationally). My dream needs a benign animal, and the playful penguins fit this shift in emotion. The perception of the precipice is also altered to a waterfall to bring it into congruity with my rapidly changing mood and animal personnel.

One week after recording this dream, I used it as the basis of a speculative model that can be easily visualized. The following table illustrates the sequence of scenes in this dream and shows what I think may be going on in my brain.

A Sequential Model of the Split Whale Dream

Event 1

Activation

Brainstem centers activate vestibular and limbic systems, which create positive feedback linking sensations of vertigo and anxiety.

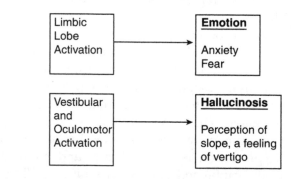

Event 2

Synthesis

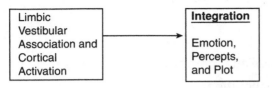

Plot Construction via Association Cortex Activation:

I am on a hillside (as at my farm in Vermont).

With my son, Ian (we are often together in that setting).

The fence and the bulls are congruent with this assumption.

So is the fence post (but not its construction).

Hippocampal and cortical networks activate and output familiar orientation data. The sense of danger continues as we see the bulls. We'd better not proceed.

Possible Memory Sources:

Cattle in Vermont

Cattle in Arizona

Event 3

Themes intensify via positive feedback and accelerating stimulus from brainstem. I explain the fence post quasi rationally with ad hoc dream logic as being composed of cedar log (for strength) and plastic pipe (to resist rotting).

Event 4

Scene 2

Anxiety is stronger, hill is steeper.

Orientational assumptions no longer fit the data so scene changes to seaside cliff and I jettison the fencepost as unnecessary (and incongruous). My son is no longer present (discontinuity).

Possible Memory Sources:

> I am interested in fencing and have just been in Arizona looking at fences, cattle, etc.

Event 5

Cliff gets more and more perilous.

Cognitive Elaboration:

> I have second thoughts about having dropped the fence post.

> Might hurt someone on beach. I am anxious about aggression.

> Plot now seems to demand some marine violence.

Event 6

Scene 3

> The split whale appears.

> Then the blubber foxhole appears.

> These items satisfy the emotional demand for violence but exonerate me. In fact, it is I who am in danger.

Possible Memory Sources:

> Steep hillsides.

> Big Sur in California one year ago.

> Mt. Kitt in Arizona two weeks ago.

> Visit to Salem Peabody Museum weekend before where I looked at whaling exhibit.

> The blubber cutting impressed me at a visceral level.

Event 7

Cognitive Elaboration:

> I decide to turn back.

> Anxiety is abating on its own or my cognitive processing has deflected and/or contained it.

Event 8

Scene 4

I see the waterfall and the penguins leaping about.

This animal is congruent with South Pacific, Peabody Museum, and plot resolution. Everything is OK, even humorous.

I bought an abandoned dairy farm in Vermont in 1965 and have since spent lots of time and money fixing it up. My most constant ally over all these years in Vermont has been my son and Split Whale dream companion, Ian Hobson. Ian is an amateur naturalist. He began by loving cows but now is an admirer of all sorts of animals, including whales and penguins. We share a love of nature. And we often tour the fields and woods of our place in Vermont together.

My interest in fences could stem from the fact that I maintain over 1 mile of barbed wire fence at my farm. While I am impressed with Vermont native know-how in putting up these structures, in keeping the wires tight, and in replacing the posts, a part of me is searching for a better system. Hence the plastic-covered cedar post! To a Freudian this post could pass for a phallus in a condom, but that's not what I think it represents. It shares with phallus/condom the preservative function, but, in inventing a new approach to fencing, it goes far beyond that function.

Many of my colleagues have hypothesized that dreaming functions in problem solving and even in creating novel devices, experiments, and concepts. Robert Louis Stevenson consulted his dream life when he was stuck for an inventive plot solution to his stories. He has avowed that the device of "taking a powder" to convert the benign Dr. Jekyl into the malignant Mr. Hyde was suggested to him by his dream brownies. In 19th century England, it was natural to assume that small creatures like dwarfs, spirits, and brownies were the mediators of such evanescent experiences as dreams. In this role, they are simply the local, cultural equivalent of the external agencies like the incubus and succubus who visited continental sleepers and gave them either bad or good dreams.

We don't yet know why some dreams are good (Dr. Jekyl) and some horrendous (Mr. Hyde), but "taking a powder" is not a bad idea. It says that a person's dreams, and even his character, have a neurochemical basis. My wife, Lia, who is a neurologist, had just told me before this dream about a patient who claimed that his behavior had been transformed by the psychoactive drug he was taking. The expectation of behavioral change was, of course, the reason for his taking the drug in the first place. But whether or not the drug relieved his daytime depression, the patient reported that, at night, he becomes a monster, attacking his dear wife in her sleep. At this point, the psychoanalyst in us clucks and says, "He has an unconscious wish to harm (or even kill) his wife."

But is this explanation fair? Or should we consider alternative theories? With the risk of raising grave problems for jurisprudence, we must ask the question of whether chemical change in the brain is the agent of all such desirable and undesirable processes. Let's examine the case for problem solving.

Fence posts *do* rot. Since the cows have left my Vermont pasture, my once tidy and tight fencerows are now a scraggly mess. Should I replace them (at considerable cost) or wait until some romantic would-be farmer comes along and does it for me? These questions are on my mind. And I know that pressure-treated posts last longer, because my treated paddock fence is still solid after ten years of exposure to the elements.

At the time I had the dream, the paddock fence was falling down, and we still had cows to keep in, and so, in due course, my dreaming brain came up with an inventive proposal: Coat the sharpened tips of raw cedar posts with plastic. As far as I know, I never had this idea in waking consciousness, so it does qualify as a novel creative product of my dreaming. But, now that I am awake, I can exercise a critical evaluation of this idea—something that was completely beyond me during the dream itself.

For starters, how could such a plastic shield be applied? By dipping the sharpened cedar posts in a big pot of epoxy? I have done this with 5-gallon cans of creosote and not seen a remarkable increase in the longevity of my posts. It could be done, but my dreaming mind never asks the how questions. Nor does my dreaming mind wonder if this approach would even work.

How would such a post stay in the ground? One of the good things about a raw cedar post is that it stays put for at least five years. Would a plastic tip be as solid? Our waking consciousness has reason to wonder. Plastic is slippery, after all. Maybe these rot-free posts will not hold. In the light of day, we say, "Let's try it."

And we could do an experiment. Just make some plastic tip coats and see what happens. But our dreaming brains don't think that way. Maybe they can't and be so freely inventive, too.

Could the plastic coating trap the moisture on which rot microorganisms thrive so that, once wet from rain or melting snow, the posts would degrade and pull right out of their plastic sleeves? I don't know the answer to this question either, but at least in waking I can ask it. In dreaming, I am as uncritical as I am unaware of my own somewhat-rotten brain condition.

After looking at the few systematic attempts to test the problem-solving hypothesis, the idea that dreams solve the problems of our waking lives, the conclusion can only be that the evidence is underwhelming. In this domain, anecdotes hold sway. Kekule's image of the snake suggesting the ring structure of benzene is said to have come to him while he was dozing on a tram. And Otto Loewi's crossed frog heart perfusion experiment—which put chemical neurotransmission on the map—is supposed to have come to him in a dream. Loewi even claims to have had this creative dream one night, and because he couldn't remember it, he vowed to have it again. And he did. We now know that the chemical that slowed his recipient frog's heart was acetylcholine released from the heart of his electrically stimulated donor frog. And my lab established that acetylcholine mediates dreaming and hence is as potent a dream powder as we can imagine.

We love such stories about dream invention because they make us feel comfortably rational in the face of overwhelming evidence that we are not. We are as crazy as coots when we dream. If we get some kind of creative mileage out of it, so much the better.

As for the bulls, we *do* occasionally encounter one. When our neighbor, Marshall Newland, lets his cows and heifers freshen by becoming pregnant, his bull goes from one cow rump to the next and can service the whole herd in a couple of days. Sex is rearing its ugly head again, but again I don't think the bulls are stuck into this dream as substitutes or symbols. They are creatures whose threat to Ian and me is a real, associatively appropriate, powerful perceptual response to fear in an open field. They also set up the sequence of animals from humpback (bulls) to whale carcass when the country scene in the field gives way to the seaside cliff over which I throw the fence post. In any case, I am often exposed to cattle and had just returned from Tucson, where I saw lots of

Western beef steers and made careful drawings of the fences and gates used to contain them.

Whales were on my mind because my first wife, Joan, and I had made a trip to the Peabody Museum the week before. I spent a good half-hour in the excellent whaling exhibit there and was particularly impressed by the methods of blubber cutting and "trying" that were so well documented. When I was 8, I was forced to spend a week in bed because I had a severe case of chicken pox. To while away the time, I read Herman Melville's *Moby Dick* in the Random House Edition, graphically illustrated by Rockwell Kent. At that early stage of my life, I had no idea about the allegorical significance of *Moby Dick* but enjoyed it thoroughly as an adventure story that documented the techniques and culture of whaling. I wanted to impress my mother, who was also an avid reader. At 600 pages it was the longest book I had ever read, and the images have stayed with me to this day.

Did Herman Melville anticipate Sigmund Freud, as many of my Wesleyan professors claimed? Was Captain Ahab the unwitting victim of a death wish that culminated with his drowning in the deep dive of the great white whale? Did his amputation represent a threat to his masculinity that could only be redressed by the vengeance of Moby Dick's capture?

Anyone who was an undergraduate at a liberal arts college in the 1950s filled up countless examination blue books with speculation about questions such as these. In the same way we are suckers for anecdotes about dreaming as creativity unleashed. We are easily taken in by the facile Freudianism of literary interpretation. We love cute stories and nod assent too readily. I read *Moby Dick* without any insight about the putative subtext. And I feel tolerant enough about the intellectual exercise provided by the application of psychoanalytic concepts to literature. But that has little to do with neuroscience.

I now return to the question of jurisprudence I raised earlier. Can sleep, with its own set of brain powders, turn a caring Dr. Jekyll by day into a murderous Mr. Hyde by night? And, if so, who is to blame? A few judges are beginning to wonder. Is a person legally and/or morally responsible for his thoughts, feelings, and actions while asleep? Most of us accept our dream life as evidence of strong and forbidden desires.

Such acceptance is easy enough if our dream scenarios are contained within our heads. But what about the patient who starts hitting his wife while asleep? Isn't he quite right to ask his neurologist if that drug he is taking has changed him, involuntarily, into a monster? And what about patients with REM sleep behavior disorder who act out their sometimes-violent dreams? One hits his wife with his left hand as he imagines making a sharp turn on the dream road. Another injures himself by diving off his bed into a dream swimming pool. Are these people responsible for their actions? Or should they be considered to be suffering from temporary insanity and therefore be pardoned and treated? These rhetorical questions are, so far, easy to answer. Patients with brain automatisms such as epilepsy or schizophrenia over which they have no control deserve understanding and such treatment as is available. And others, especially spouses, should be protected from them.

But once a "sleep defense" is admitted into court, the picture can rapidly become considerably more muddy and much more troubling. Consider the young Canadian who turned himself into the police when he heard of a media report of the murder of his in-laws. He wondered if it was possible that he was the murderer and if he could have committed this crime while he was asleep. The judge, jury, and medical expert witnesses all said yes, it was possible that a person could get up from his bed, drive 30 miles in the wee hours of the morning, murder his parents-in-law, and drive home without being truly awake. Not guilty by reason of sleep disorder was the verdict. Does this verdict strain your credulity? Beware! Jurisprudence is based on precedence; your child might marry a somnambulist; you yourself might have to take a powder that changes your very being.

In the Canadian case a motive was evident. The in-laws felt that the sleepwalker/murderer was an unsuitable match for their daughter. When she married him anyway, the parents did not hide their displeasure. The hurt feelings smoldered and were made worse when the suspect lost his job and suffered a major loss of self-esteem just a week before the murder. But many people have at least as strong a motive and don't commit murder. What went wrong here? We don't really know, but the sleep experts who recorded the suspect's sleep in their laboratories told the judge and jury that the suspect had dissociative states in which he, like one who has been hypnotized, was partially awake and partially asleep.

Subjects with dissociative states have been called "somnambules"—meaning, literally, sleepwalkers—since the time of Pierre Janet, who was Jean Martin Charcôt's heir apparent at the Salpetriere at the time of Sigmund Freud's famous visit. And classic sleepwalkers are certainly dissociated. Between the ages of 9 and 11, I am told, I would get up, walk down the stairs, open the front door, and urinate in the rose garden. Was I motivated? Yes. I needed to pee. Did I know what I was doing? No. I was in a dissociated state with some elements of waking and some of sleep.

The historic tie of sleep to dissociation was broken when Freud put his emphasis on motivation and ignored the process of dissociation itself. Now the story has come full circle and sleep can be seen, by its very nature, to provide the physiological ground of a wide range of motivated but uncontrolled behaviors. What will the legal system do with this new way of seeing things? The answer is good news for lawyers, whatever the final practical verdict may be. Will the drug companies that produce behavior-altering drugs or the doctors who prescribe them be sued for damages by victims of dissociational misbehavior? It seems inevitable.

We are heading toward a new view of the unconscious. The theory that dissociative states are caused only by the pressure of unconscious wishes lies at the heart of Freud's theory. Dreams, and all neurotic behavior, were to be understood in the same way—as deformations of conscious experience caused by the eruption of repressed ideas and impulses. Janet and Charcôt were not so sure about this suggestion of Freud's. While they agreed about the potency of repressed impulses, especially sexual impulses, they sought but could not find a neurological substrate for the symptoms they observed.

Now we have it. The brain changes state so dramatically in sleep that we are not always the same people in dreams as we are in waking. What is crucial here is the concept of state, and that is what modern sleep research is all about. We now know that it is natural, normal, and physiological to have dissociative experiences when falling asleep, when waking up from deep sleep, and when waking up from dreaming. This is a truly revolutionary fact, only beginning to sink into the modern psychiatric mind.

At sleep onset, we are frequently both awake and asleep. The most striking—and dangerous—situation illustrating this point is falling asleep at the wheel. Potentially fatal lapses are often preceded by suspensions of conscious awareness called "highway hypnosis." This term honors the myriad hypnoid states, including formal hypnosis itself, which illustrates the brain-mind's capacity to be there and not be there at the same time.

Once sleep makes its demands more exigent, we are prone to visual illusions, and even frank hallucinations, as we struggle to stay awake. These borderline states have as yet received little attention from psychophysiology because surface EEG recording is not sensitive enough to objectify the changes in brain function that underlie them. PET and MRI are useful in explaining the dynamic neurology of these states. But we already know that attentional lapses and endogenous percept generation do occur when the EEG hovers between waking and Stage I sleep.

The point here is not to discount the role of motives in unwanted actions but to emphasize instead the importance of the changes in brain physiology that provide the necessary substrates for dissociative behavior.

Another example that is relevant to understanding our young Canadian's murderous adventure is the difficulty of changing our state to waking once sleep has established itself. We have all experienced what cognitive scientists call state carryover phenomena. We are behaviorally awake, all right. Witness our capacity to shave, brush our teeth, and shower when the alarm goes off at 5:00 a.m., signaling our need to get up early. But our mind is still foggy, often for several minutes, as our brain remains partially committed to sleep. The same process is even more powerful earlier in the night. If we are called by a friend in need, or by a patient, or simply by nature, it is often quite difficult to respond because we are so deeply asleep. In these dissociated states, the EEG might show that our upper brain is still deeply asleep while, by inference, our wake-like behavior is entirely governed by the more-alert lower brain. This full-blown somnambulism, better known as sleepwalking, makes our point crystal clear: One part of us is awake and one part is still deeply asleep.

Such frank and complete dissociations are so common in early adolescence as to demand that we look at them as normal phenomena. Once we cross this bridge, it is much easier for us to acknowledge that the many more-subtle forms of dissociation, assigned by Freud to the category of psychodynamically determined neurosis, are the entirely normal physiologically and inevitable properties of the brain's own dynamism as it changes state. To make my point clear, I call such phenomena neurodynamic. Motives play their part in some of the more flagrant cases, but the brain has its own laws for changing state, and these must be distinguished from the psychological constraint of unconscious wishes. The brain is a complex physical structure that has its own dynamic regulatory mechanisms. It enters the REM state whatever our motives might be. When it enters REM, we dream.

Dreaming is the dissociated state par excellence, and our new scientific understanding allows us to distance ourselves from the psychoanalytic view of it. Here is how our alternative explanation goes: When we dream in REM sleep, our brain is activated. It thinks it is awake. And we think we are awake. And it is true: Part of us is "awake" in the sense that our brain, in most of its parts, is as turned on as it ever is when we are really awake. No wonder we are so constantly fooled. We are in a sleep state with many features of waking. REM sleep dreaming thus is deeply dissociative, having two properties at once: simulated waking and sleep.

Sometimes we wake up from a dream unable to move. In this variation on the theme of dissociation, the motor paralysis of REM extends into waking. Narcoleptic patients, or people taking antidepressants, might have disturbing visual illusions or even frank hallucinations after they wake up. A part of their brain is still in REM sleep, and another part is asleep. Dissociation is again at work.

Given the complexity of the brain and the necessity to coordinate all of its 100 billion neurones so that they are in the same state at the same time, it is truly wondrous that we suffer so little from this built-in propensity to dissociate. It is not so much neurosis that plagues us, it is possessing a brain over which "we" (a construct of our brain) can exercise only very partial control.

The ecological theme of my Split Whale dream is latent but clear. I am always aware of the risks and threats of my encounters with animals, be they bulls or whales. Both man and beast stand to lose (dignity, life, and diversity) as well as to gain (money, milk, and oil) from our encounters with them. Ian is a part of this complex because he feels for animals but also because he himself is handicapped by mild congenital brain damage.

This dream's happy ending is unusual! More frequently I would awaken at the end of scene 2 or 3 and never know that playful penguins were just around the corner! Why does my Split Whale dream—so frightening and menacing through most of its course—suddenly become playful, amusing, and pleasant? We have no idea what causes such abrupt shifts in dream emotion, but we suppose that it has to do with activation of the part of the limbic brain that mediates positive emotion. We know, from Ritchie Davidson's work, that positive emotion in waking is associated with activation of the left frontal cortex (and the left amygdala), while negative emotion is associated with right-sided activation. If this rule applies in sleep, how is the shift affected? We want to know the answer to this question not only for the sake of completeness of our dream theory but also because we would like to understand how positive emotions can be enhanced via cognitive interventions.

A cogent example is related to our intent to help people gain some degree of control over their nightmares and "bad" dreams. Although REM sleep is entirely involuntary and involves highly automated physiology, it is possible to introduce a significant degree of voluntary control via lucidity training. In sleep, as in life generally, a little bit of volition goes a long way.

I have set out my ideas about dream lucidity and indicated that cognitive priming can increase our chances of recognizing, via dream bizarreness, that we are in the dream state. As a variation on the theme of dream hedonism, it is also possible to change the emotional tone of dreams. Recognizing the negative emotion as such enables the nightmare-enmeshed dreamer to escape by saying, "Get out of my dream; I don't want to be frightened." Seeing if we can arrange for penguins to slide down the waterfall is not the reason that my Split Whale dream had a happy ending—but it might be.

The settings, animals, and objects have an associative link to one another. Hillsides, cliffs, and waterfalls are all vertical geographies. Bulls and cows are not just the names of farm animals. They apply to whales as well. And penguins fit with whales as exotic overseas creatures. Finally, the form of the fence post matches the form of the whale carcass. This dream is driven by the strong emotions of fear and anxiety, and it ties together a set of clearly identifiable themes. At first glance, farms and whales have little apparent relationship to one another, but in terms of a dream logic that is based on shared emotional salience and experiential similarity, they fit quite nicely together.

Chapter 10

Tiffany Box
Toward a New Philosophy of Mind

I dream a long dream and wake up at 5:00 a.m. A glass box of the Tiffany period is also, somehow, a room. At first (after several unremembered scenes) it is perched on the back of a truck, which I must maneuver as close as possible to the wall edge of a bridge-dam. To my passengers I say, "Don't worry," and I zigzag perilously close to the edge.

Now we have passed security (in a university building?) and emerge on a rocky promontory above the waterfall, which we (as a group) are to plummet down. The idea is that we will all get in the box (which suddenly seems incredibly heavy) and—at the same time—ease it over the edge. I have already scaled these same falls alone (with exhilaration) in a bathing suit. The impossibility of lifting the box while being inside it dawns on me, and I decide to investigate the falls, which seem more perilous than expected.

I walk to the near precipice and look down. There's no way to survive the rocky staircase below. The shattering of the glass menaces. (David Armstrong has said that glass, being brittle, has the intention to break.) Below and to the left is a long, vertical, unbroken waterfall, which, I suddenly realize, is the way I went before. The sense of three-dimensional depth is strong, with a huge, deep gulf that beckons and forbids at once.

I decide it is unsafe (as well as impossible) for us to go over the falls.

The sense of foreboding—and transgression—increases. It is clear that the police (or militia) are coming. They are Communist guerilla types in fatigues. A platoon of about 20 swarm on to the plateau and surround us.

The leader, who speaks to me in friendly, condescending, even obsequious tones, is young, smiling, and has a remarkable moustache.

It is only 1/4-inch wide, but the hairs stand out at least 2 inches and form an ogee curve around his cheeks. A tall, thin, ultra-walrus moustache increases the leader's comic dignity. It makes him look a bit like a punk-rock Fidel Castro.

Speaking for the group, I decline the suggestion that we join the Communist party and, when the anticipated threat fails to materialize, I wake up.

This is a "Mission Impossible" dream, the kind of dream that typically involves perilous heights that must be climbed or mastered in some way. Sleeping at home, I awakened at 5:00 a.m. with excellent recall of this very long, animated, and complex dream. As with other early-morning recall, I am certain that there were preceding scenes I cannot remember.

The dream occurred on the night after David Armstrong, the Australian philosopher of mind, had given a seminar in my Brookline living room. Armstrong's views are decidedly materialist, so we are in agreement on many neuroscience questions. Materialist philosophers, like David, are pleased with the burgeoning growth of neuroscience and with its growing capacity to contribute to the philosophical debate about the mind.

When first I met Armstrong (at the St. Andrews, Scotland conference stimulated by Adolf Grundbaum's critique of Freud), he had reassured me about the value of my presentation to this highly technical philosophical group. "You have more interesting data than any of us," he said.

Besides Armstrong, another apparent memory source is my psi drawing, which becomes inverted and slightly altered as part of the box design. I am an amateur artist-designer-architect, and I often render my ideas visually. But it is unusual for them to make their way into my dreams either before or after the fact. This dream is powered by the emotions of anxiety and elation. The visual images are lush and detailed. From the Tiffany box, through the Niagara-like falls, to the Communist platoon leader, I could see every element clearly. There was also continuous movement. My sense of vertigo and the three-dimensionality of the void must have been inspired by strong visuomotor and vestibulo-cerebellar activation. Since having a stroke in 2001, which damaged my vestibular cerebellar system, I don't have these vertiginous dreams anymore. In a way, that's a relief, but I don't fly, either. I regret that. The images are so powerful that I can neither deny them (and realize that I am dreaming) nor use my head (to plan a strategy for my leadership behavior). My reasoning power is deficient, and I go from one impossible dream idea to the next.

This dream affords excellent illustrations of the defective thinking of the dreaming brain. There are two aspects of note: the absence of reason when it would be expected, and the weakness of reasoning when it does occur. A glaring example of absent thinking is my failure to recognize the marked improbability of a glass box of the Tiffany period that is also, somehow, a room. Had I seen such a box in waking, I would have been perplexed, even if I were in an art gallery. Such a box is not only intrinsically impossible but, in the dream, is incongruously set outdoors in what will soon becomes a wilderness setting. The best I can do to reconcile these incongruities is to place the box-room on the back of a pickup truck. But this transmogrification is accomplished without any conscious thought or cognitive effort. It just happens—entirely within the perceptual domain.

I am aware of the truck's perilous position (on the wall edge of a bridge dam), so I therefore reassure my passengers. "Don't worry," I say. But they should worry. And so should I. Things are happening too fast for me to exercise even weak reasoning, such as "Of course, you are right to be worried. How the hell did we get into this situation in the first place? I have no recollection of having driven out here, so I must either have amnesia or be dreaming." But no such thoughts enter my dreaming brain.

The following scene is even more absurd. How in the world can we lift a box if we are all standing inside it? And why in the world should we want to ease it over the edge of a precipice? And where did that waterfall come from? What happened to the bridge-dam? I finally recognize the physical impossibility of the lifting task (pulling one's feet up by one's own bootstraps), but the best I can do to fix this crazy situation is to investigate the falls, which appear perilous.

At no point do I consider calling off the foolish mission or turning back, as even an impulsive guy like me would do in waking. No, no, it's full speed ahead and damn the torpedoes, as I lead my crew to almost-certain disaster. The point is that, like a manic patient, I have little insight into the impossibility and danger of the cockamamie project. I do decide not to go over the falls at point A, but it takes a long time and a mountain of evidence to force me to this point.

It is, by the way, during a lull in the dream that I finally reason my way out of peril. Or nearly so.

Having decided not to proceed over the falls, I am suddenly confronted by the Communist guerilla platoon. This actually continues the emotional theme of fear and menace but shifts the blame for it from my shoulders. I now ascribe it to an imaginary enemy, which makes the physiologist in me wonder—was this scene change caused by a sudden decrease in the intensity of eye movements and associated limbic lobe activation?

Such a process could allow the brain to cool off a bit, as well as set the stage for a scene shift. Indeed, the driving emotion becomes much less intense and, after the shift from me and my ill-fated project to the guerillas as a source of fear, the emotion actually shifts from negative to faintly positive as the guerilla commander with the walrus moustache appears, amusing me a little.

But I am still not thinking at all critically. How could there, and why should there, be Communist guerillas lurking in any woods or mountains that I might adventure in? My mind cannot see that. Shall we emphasize the power of the emotional and the visual or the weakness of the rational and mnemonic minds? Or both? I think it is both, and the positive and negative forces are not only additive but multiplicative.

In other words, my perceptual-emotional experience is at once incompatible with and enhanced by my cognitive deficiencies. Without memory and critical thought, I am completely at the mercy of what I see and feel. Likewise, the intensity of my perceptual-emotional experience preempts memory and reasoning. To account for these autonomous aspects of dreaming, there must be, as Wilhelm Wundt opined 150 years ago, a brain process that is enhanced and another that is impaired. The enhancement, I suggest, is cholinergic pontine-limbic associative cortex activation. And the impairment is the aminergic demodulation of the frontal and prefrontal cortices. It happens every night like clockwork.

Memory, even remote memory, is conspicuously weak in this dream. No specific place, time, or person is specified. If David Armstrong's visit, seminar, and paper have something to do with the content, why doesn't he (or his wife) appear in the dream? And if the falls have something to do with Frederic Church's or my childhood images of Niagara, why are those details not specified? And who is this mustachioed platoon leader? After the fact—

and only after the fact—I suggest a punk-rock update of Fidel Castro, but in the dream I don't know this guy or what his name might be.

The Tiffany period box that is also a room is a classic dream mystery object allowing associations to proceed in any direction. In that limited sense it is, I suppose, a symbol. But in my opinion it isn't a symbol of any one thing. Should we assume that it is a mental product designed to conceal a specific motive in its ambiguity? Or is it rather an artifact of brain activation in sleep that is designed to carry several meanings, none of which is as important as the emotional salience of the dream's main theme, which is perilous navigation through a menacing physical and social environment?

The best I can do with the box itself is to note that the back doors have a form resembling the Ψ (psi) logo that I have proposed for the Massachusetts Mental Health Center's new image campaign. Approaching age 75, MMHC is teetering—like the truck—on the brink of disaster. It has lost its identity, its sponsorship, and its way of being. But I can see no reason to symbolize my fears of institutional extinction, which are quite conscious and manageable. I do my best to help, but I feel that my wake-state scientific mission—which is ambitious but not impossible—must remain my top priority.

Having transformed the otherwise-useless box into the black body of a pickup truck, I begin to maneuver the vehicle (my laboratory?) along its course while managing my passengers (my colleagues who worry about our future?) by saying "Don't worry" as I carry out a perilous zigzag along the edge of the dream bridge-dam. As a margin note in my journal, I have written, "I have been admiring Toyota 4×4s and pickups," probably because my daughter Julia, age 13, had said she loves them and because both she and her wayward mother want to spend more time in the country pursuing horseback riding and related romances. Maybe it was also because the price of gas was coming down at the time.

The setting suddenly includes a distinctly incongruous security checkpoint. But instead of being safely inside, we are perilously outside on a rocky promontory above a waterfall that we must descend. It's a bit like the Niagara Falls fantasy that has had a place in my mind since I went under the Canadian Falls on the

Maid of the Mist at age 6 and learned that some intrepid fools went over those same falls in a barrel.

How can we proceed? Dream reasoning is called to the rescue but, as we have already seen, it is a fool's errand. At first I think we will all get in the box (no wonder it gets heavy) and ease it over the edge. Only later do I realize that this is physically impossible. Reference to rational thinking is two times less frequent in REM than in NREM dreams, and the hallucinatory drive in REM is reciprocally greater.

It is equally unlikely that I could have gone down this perilous route alone in a bathing suit, but I do not consider that possibility. Instead I recall my exhilaration, which is typical of the death-defying and physically impossible flying dreams we had recently been discussing in my Introduction to Psychiatry class, as my dream journal notes. Once I realize the descent is impossible, as is typical in dreams, I invent two possible descent routes where one would do.

In a margin note in my dream journal, I emphasized the site's physical appearance, which is reminiscent of Frederic Church's painting of Niagara, but narrower and more gorge-like. From the box location on the precipice, I see a rocky streambed and imagine the glass walls of the box shattering into smithereens.

To help explain my earlier misjudgment about a possible descent route, I then behold a second, long, unbroken vertical cataract that simultaneously evokes a pull (if I could fly, I could go down this way) and a dread (I know it is impossible). Only then do I decide that descent is impossible.

Then the scene changes! But the feeling of dread does not dissipate. Instead, the sense of danger and poor judgment increases. To match this feeling I select a political/police scenario that is not common in my dreams and that might respond to Armstrong's essay on Poland, which I have just read. But the Communist guerillas and their mustachioed leader came from nowhere in particular that I can identify. Another symbol? Military authority? Blind authority that I wish to mock? A proxy for the state, which is deconstructing the Massachusetts Mental Health Center? All these are possible sources, but none can explain the scene better than a shift in my mood from fearful apprehension to elated confidence.

Note the comical composition of the final scene. Even the most preposterous walrus mustaches don't stand out 2 inches from a 1/4-inch-wide base! So it is not I who is ridiculous for trying to get a Tiffany box full of people over a waterfall, it is he who expects me to deliver my group to the Communist party. This is an easy request to decline, and I do so with impunity.

The neuroscience of sleep and dreaming is stirring up philosophers. For a materialist, like Armstrong, it is powerfully reinforcing of his monistic inclinations. For him, the mind is inseparable from the brain and, when we know enough about the brain-mind, will be reducible to it. That doesn't mean that mind, consciousness, free will, and those feelings of awe that we all cherish will go away. For better or worse, subjectivity will still be with us, and it will still be what we experience. But we will understand it better, says Armstrong. David Chalmers, also an Australian, calls the brain-mind connection the "hard problem" of philosophy. How can a material thing, the brain, which is made of jelly and whatnot, be capable of subjective experience? he asks. No matter how much we learn about the brain, we will never be any wiser about subjectivity.

But we are already considerably wiser than we were in the century since Freud, and in the last fifty years the pace at which we are growing wiser has clearly been accelerating. What has changed is our physical knowledge of the brain. Philosophy has to run to catch up and, to be fair to Chalmers, he is running hard. Since the quality of conscious experience changes in such a stereotypical and predictable way when the brain changes state, there can no longer be any reasonable doubt about the brain basis of consciousness. It may be another century or two before we have an adequate answer to the "exactly how?" question, but you don't have to be a weatherman to see which way the wind is blowing.

Like David Chalmers, we often ask ourselves how subjectivity could possibly arise from a physical object like the brain. But this question can be turned on its head. We could ask, how can the brain not be conscious when it is composed of over 100 billion neurones, each connected to 10,000 others (and add firing at rates of 2–60/sec)? The amount of information processed is upwards of 10^{29} bits per second. That is enough information to support both perception and the perception of perception and enough to support awareness and awareness of awareness. In a word, it's enough to support consciousness. So it is possible. Now the question becomes, how is it possible? Not whether it is possible.

What sleep and dream research adds to the picture is the parallel analysis of state changes in the mind and state changes in the brain. The fact that they go hand in hand means either that the connection is one-to-one or that they are somehow one. Getting to the details of how this can be is now a feasible project. Dreaming, in fact, provides a very good primer for any naturalistic philosophy of mind. Contemporary scientific philosophers such as Patricia Churchland in San Diego and Owen Flanagan at Duke University in North Carolina are well aware of the impact of neuroscience on the philosophy of mind. In dreaming, a whole world of conscious experience opens up when the brain is offline. This means that, for the time being, anyway, the brain mind uses its own energy and its own information to create a complex and exciting virtual reality.

Our discussion need not be limited to living philosophers. The results of the neuroscience revolution favor Aristotle over Plato, Leibniz over Locke, Kant over Descartes, and Chomsky over Piaget. For Plato and Descartes, the blow is to the assumption of the primacy of ideas and, more specifically, to the independence of ideas (and the ideal) from material substance and the real. Just as there can be no noise without a hearing ear, there can be no thought without an activated brain.

Aristotle was the first great biologist and the progenitor of observational experiments that have a couple of millennia later led us to know so much about the brain. Leibniz argued that the brain—and the mind—continued to be active in sleep. He used the continuity of consciousness, across nights of sleep, as evidence of what he called *les petites perceptions*, which were the proposed basis of such continuity. In brief, Leibniz was correct in his intuition of continuous brain-mind activity. Locke's idea of the brain as a blank slate or "tabula rasa" on which the environment inscribed instructions is obviously wrong.

The fact that brain activation precedes birth and is dramatically present in early development strikes a blow to Descartes and Piaget and favors Kant and Chomsky with their notion of structural and functional competence preceding experience. There is no doubt that experience matters, but experience interacts with an already-elaborate structure-function—the activated brain-mind. So the categories of experience precede thought, and grammar precedes language.

For all dualists, the discovery of brain activation as an antecedent to thought and dreaming is a mortal blow. Mind and brain are not two mechanisms set to run from birth to death on parallel tracks. The physical impossibility of such a theory is guaranteed by the unpredictability of the brain-mind, which, like all complex systems, has a built-in capacity for chaos. If the brain and mind were not causally linked, this capacity alone would make synchronization even more unlikely than if the two domains were interdependent.

God is very clever, but this is a bit too much to ask, even of him! Why not instead create a brain and have consciousness emerge when that brain reaches a sufficiently high level of organization?

Chapter 11

Louis Kane Dies
The Brain Plays Tricks with Memory

I have a 10:45 a.m. appointment to meet my old friend Louis Kane at the Harvard Club, where we will play squash. I walk into the large, broad lobby and see the usual preppy-looking business-suited types, but there is no sign of Louis. This begins to disturb me at about 11:00, because he is usually punctual, especially for squash, for which we must reserve a court. When I ask some unidentified club members if they have seen him, a black-suited porter-valet type announces that Louis has just been found, dead, in his room. The valet is upset, because he knows Mr. Kane well and admired him deeply, as almost everyone did.

I am surprised by this news but not particularly upset because Louis has been so sick for so long that his death seemed quite likely. I ask the valet if the family has been notified and offer to do so myself. "His sister and father are staying in a hotel around the corner," I am informed. Not recognizing the hotel's name or location, I ask twice and am twice given the requested parameters. But when I set out to find the hotel, I realize I am completely lost. Furthermore, I have forgotten the hotel's name and address. This makes me very anxious, and I wake up.

My new wife, Lia, was on duty at the hospital in Messina, Sicily. I went to bed at 10:00 with the twins, who had just finished the third *Harry Potter* book. There was a nice breeze coursing through our apartment, so I slept well until 2:30 a.m. when I awoke, dismayed to find that Lia was not yet home. When she finally returned, after a late Sicilian summer soiree with three of her lady friends, I went back to sleep. Sometime between then and 4:00 a.m., I had the dream, which I remember well. I lay awake thinking about how typically strange it was. I recorded this account first thing in the morning. The fact that it is about death makes it especially relevant.

The dream falls into a particular category: Someone, often a close friend, who is already dead, is the focus of dream attention as if he were still alive. In this case, my friend, Louis Kane, dies for a second time and, during the dream, I have no access to the information about his first death. This information is immediately available to me when I wake up. Thus, the dream illustrates the remarkable dissociation of narrative memory that is typical of all dreams, not just reincarnation dreams like this one.

Louis Kane died on his porch in Maine a month before this dream. He had battled pancreatic cancer for several years. Recognizing that he was losing the fight, he had decided to go to Ogunquit to be with his family in the place he loved best. Sensing what was coming, I drove up to see him in June before leaving for Sicily. We didn't actually say goodbye, but we knew that it was unlikely we would see each other again.

None of this highly salient memory material was available to me in my dream. Psychological overdetermination (as in "I wish he were still alive") is not the answer modern neuroscience would suggest. It is certainly true that I wish he were still alive, but there is no reason for that wish to be unconscious or efficacious in walling off memory material that is so readily accessible during waking. The answer must be that access to narrative memory is physiologically impeded during REM sleep dreams, as Bob Stickgold's theory states.

Other anomalies support this answer. One is that the lobby of my dream Harvard Club bears no resemblance to the Massachusetts Avenue Boston Harvard Club where Louis and I used to meet to play squash. It doesn't even resemble the New York Harvard Club, which I can almost as easily visualize in my waking imagination. In my dream, I enter instead an ersatz Activation-Synthesis Harvard Club gotten up to form a quasi-legitimate setting for my dream meeting with Louis. New York comes to mind because it is there, and certainly not in Boston, that I might have found Louis's sister, Annie, and his father, George, in a hotel. In Boston, they live at home, and I know perfectly well where their homes are. But what hotel did my dreaming brain assign them to? And where, exactly, was it? My dreaming brain is as incapable of getting a fix on this orientating information as it is creative in conjuring up novel architectures, settings, and characters.

It took me and my fellow researchers five years of research, defining and measuring dream bizarreness, to recognize what was going on. Consider the flagrant disorientation of this dream. In waking, I know perfectly well what the Boston Harvard Club looks like. Even though I have not been there in over twenty years, I daresay I could draw a reasonably good floor plan. And although I am a poor if enthusiastic artist, I could even draw sketches of the lobby, the newspaper room, the stairway down to the squash courts, and the billiard room that we crossed on our way to the locker room where we changed before playing squash.

All this information, and more, is in my brain, but in sleep I don't have access to it. Freud had no other way to account for such inaccessibility other than his invented mechanism of repression. But as this and many of my other dreams make clear, repression is not the only way to compromise memory, and it might not play any role whatsoever in dream memory deformation. Dream memory failure is more likely a form of functional but entirely organic amnesia.

Memory, like so many other psychological processes, is state-dependent. The state of the brain changes dramatically in sleep. And so, necessarily, does memory. In Freud's theory, amnesia was thought to be psychodynamically determined if there was no structural organic brain damage. Charcot, Janet, and Freud were all well-trained neurologists who wanted to explain the defects of memory shown by their "hysterical" patients neurologically. But they could not because they did not have the concept of dynamic neurological dysfunction. For them, as with most Freud followers, amnesia in the absence of structural neurological disease had to be ascribed to something like repression. This error now appears as one of the greatest gaffes in the history of ideas, because it has misguided our understanding of who we are and how our brains work for over a century.

My Louis Kane Dies dream provides a neat refutation of Freud's notion of repression. There is no reason for me to repress architectural details of such things as the décor and layout of the Harvard Club. Rather, because of a functional change in memory access, I cannot summon, while dreaming, orientational details that easily spring to mind once I am awake. The "secret" of dreams is nothing more or less than a functional change in brain function during sleep.

Time is also warped. I have not played squash with Louis for at least twenty years. In fact, I haven't played squash with anyone for more than ten! And yet it seems quite normal to me, in the dream, to be tuning up for a squash date with Louis. However anachronistic, my squash readiness is emotionally salient because we so much loved playing together, sitting in the steam bath afterwards, and sharing stories of our lives. If it were operative, a wish-fulfillment motive would have surely supplied me with the pleasure of a game and an aprés game. Twenty years ago, neither of us thought much about dying. But today, he is gone.

My Louis Kane Dies dream is not particularly animated. I go to the Club and meet the porter who informs me of Louis's death, but I do not actually play squash. In fact, I believe that there is not a single squash-playing dream in my collection of over 300 reports. The paucity of movement in this particular dream suggests that it may have been a late-night NREM sleep event. And the absence of squash dreams in my collection could be explained by the fact that most of the reports were recorded after I stopped playing squash.

But that doesn't explain why other of my dreams do involve sports I have long since given up or never indulged in at all. Most of my wild motoric dreams involve individual action whether they are sports or not. Hence my Caravaggio dream involves biking (which I was still doing when I had that dream) and unicycling (which I have never done). Both of those actions were in the service of social behavior, but neither was a sport.

While we now know enough to hypothesize that motoric dreaming is related to motor program activation and that motor program activation is most intense in REM sleep, we still have no more idea why certain motor programs and not others are activated than why limbic brain activation leads to some emotional experiences and not others. In other words, we still have a lot to learn. But we now have the concepts and techniques that can help us learn.

Of all the pleasures that Louis and I shared in our exuberant youth, the only one left is the joy of eating. The bodily pleasure of my dream on July 11 bears some relationship to the memory of eating and drinking well in our regular July 4th celebration in Maine. Now that Lia and my new family travel to Sicily each summer, we miss that event. Life with and without Louis had been on my mind. The relative rarity of hedonistic dreams is lamentable. When I have one, I am as grateful to my dreaming brain as I am to the fates who give me pleasure during waking. We take daily pleasure at the table; we enjoy aperitivi (in Italy) and cocktails (in the U.S.) before the evening meal. Yet this is not something people tend to dream about much.

My Bicentennial Wine dream might appear to be an exception. But although it is ostensibly about wine, I never get to taste the bottle of 1779 claret. In my mind's taste bud, I can clearly remember the palatal pleasure of wines that I enjoyed with Louis, who had a wonderful collection of great red wines. He always served good ones on July 4th. Awake, I can imagine the taste of wines such as the Cos d'Estournel and Lynch-Bages of 1975.

I simply don't eat and drink in my dreams. So many banal daily life activities, such as writing books, reading books, and talking with students, are absent from my dreams. I am sad to say I don't have an explanation for this. Emotional salience is strong in both cases. Perhaps we don't need to relearn these behaviors because they are so robustly programmed in our neurons. Pain is underrepresented in dreams because it is difficult to simulate. The same argument works fairly well for taste and smell. But this can't account for the lack of daily work activities. They are easy to simulate. This is one of the innocuous yet great mysteries of the mind, which neuroscience may be on the verge of solving.

I have always been impressed with the closeness of Louis's sister and father. They care for each other on a daily basis, and that is a contrast to the habits of my own family. The dream shows that I am still looking for a way to make my family closer. It is natural that dreams should concern themselves with projects as-yet uncompleted. In this category are all the traditional "working through" tasks of interest to contemporary psychiatry. Indeed, prominent dream theorists such as Rosalind Cartwright hold that dreams not only reveal the mind's attempt to come to terms with trauma and loss but also constitute a necessary and sufficient mechanism for psychological healing. In other words, if you don't dream about the difficulties of, say, divorce, you will not be able to get beyond it.

My Louis Kane Dies dream is triggered by my own current preoccupation with death and with missing specific aspects of life that I shared with Louis. That I should have him die again at the Harvard Club we so enjoyed is certainly significant in terms of both nostalgia and the working through of his loss. Do I really think I need to have such dreams to be nostalgic and work through loss? No. But I cannot be as sure of this conclusion as I am of the dream memory theme. Narrative memory is physiologically blocked in REM sleep. This is probably because the dorsolateral prefrontal cortex is underactivated and the two-way flow of information in and out of the hippocampus is lost in REM sleep.

When I am awake, most of the information that is stored in my brain is, despite my 71 years and a major neurological insult, readily available to my consciousness. Even today, four years later, I can still see Louis sitting on his bedroom porch in Ogunquit, looking tired, thin, and sleepy, but still vivid in his contact with an old friend. The dappled shade of the trees behind his house danced across his porch. The fine weather only made me sadder to see that Louis was fading fast.

If I were writing a work of fiction, it would be a relatively simple matter to disguise Louis's identity, give him a different name, and let this fictional persona tell the world the secrets he shared with me. I speculate that really great storytellers can summon such details from memory and recombine them in ways that are every bit as salient as to the person describing them. Dreaming may help us recognize such salience, but only the waking brain can present it to someone else in a cogent manner.

Chapter 12

Medieval Town

Bizarre Architecture, Emotional Salience, and the Healing Medulla

The action takes place in a foreign country, which I think might be Yugoslavia or Hungary. Lia and I are on a trip, and we are crossing a bridge with a lot of other people. It is a high arched bridge of a medieval sort. The bridge suddenly separates from the land and becomes a boat skimming across a very small river to get to a village on the other side. We plan to stay in an old-fashioned inn.

There is already some discomfort and difficulty finding each other as we get the boat near the shore. I catch glimpses of Lia. She is talking to someone else, a man. At one point, either before or just after we get off the boat, I notice that she has given or sold him a half-inch bit used with my large brace to drill holes in wood in Vermont. I am very surprised and somewhat hurt by this. I also notice that the bit has been used to make a perfect hole in the shoulder bag the man is wearing. It is a shoulder bag very much like mine.

Lia explains that she has sold the drill but will give me the money. It still seems odd that she would give a stranger one of my most precious tools without asking me. I feel very vexed and apprehensive. When we get on the land, we walk around looking for the inn and are separated from each other on several occasions. During one of the times we are together, she makes it clear to me that she needs to have a secret life.

When I ask her about this man, it is clear that she means she needs to be free to have an affair with him if she wants to. I find that very odd and disquieting and try to express my concerns. When we finally get to what appears to be the inn, there is a strange scene in which she is again difficult to find. But I find her in what looks like a kitchen, and she is preparing to cook some food, which strikes me as odd, since this is such a flimsy excuse. I ask her when she will be finished, and she looks at her watch and says 45 minutes, to which I agree, knowing that this was all the time she would need to make love with whichever stranger she had selected.

I then walk around the inn, which has a very peculiar structure. On one side is a row of seats, which are roofed over, as in a theater. Beside each seat is an exotic bouquet of flowers. I walk from one level of this improbable architecture to the next, finally getting to the bottom. I wonder in which room Lia and her lover are situated and how I can get to the window to see them. But by going down, I am going down from the bedroom level, as if to avoid my curiosity. I wander all the way around the other side of the building, admiring the ancient medieval architecture, all of which is quite exotic and entirely impossible. When I come back to the place where I think I might find Lia, I see her coat, the brown coat with the fur trim and hood that she wears so often and that I like so much, but there's no sign of her and no sign of whatever man she might be with.

On a recent visit to Monaco with my second wife, Lia, and two friends from Messina, I had the sudden onset of an unpleasant spinning sensation. I put my head down on the table of the Monte Carlo Casino Café, where we were enjoying breakfast, and waited for this unwelcome gyration to subside. I also noticed sweating on the right side of my face. This, plus the vertigo, made Lia (who is a neurologist) think that I was probably having a stroke. I was.

I insisted on walking back to our room in the Hermitage Hotel, where I was sure I could sleep it off. I couldn't. Sleep helped—and it is still the best anodyne to my residual stroke symptoms—but the symptoms didn't go away then, and they don't go away now. In fact, on the morning after my Casino Café dizziness attack and a very good night of sleep, I found it more difficult to walk and impossible to swallow fast enough to clear the back of my mouth of the sea of saliva that constantly welled up and made me fear drowning in my own secretions.

I was taken to the Princess Grace Hospital, where an electrocardiogram revealed an irregular heart rhythm. I still have this atrial fibrillation problem and must take a blood thinner, Coumadin, to reduce the risk of more brain damage due to the plugging of my brain blood vessels by tiny clots spun off the wall of my heart. Most of my doctors think that's what happened to me.

After three weeks of good nursing care in Monaco, I still couldn't walk or swallow very well but was well enough to be flown back to Boston in a Learjet Air Ambulance. At Brigham and Women's Hospital, the nursing care was not as good as in Monaco, and all the many lab tests could do was confirm the diagnosis of stroke due to plugging of a small artery in my brain stem while I was in Monaco.

The dramatic changes in my sleep and dreams that occurred after my stroke became the subject of several technical articles to which I can only allude here to make the point that for several weeks I missed my normal dreaming and knew that its return would signal a major and welcome advance in my brain health. After my stroke, my first elaborate and sustained dream occurred on day 38, when I was already beginning to walk and exercise at the Spaulding Rehabilitation Hospital. I had been waiting in vain for a vivid dream for over five weeks. I was in the hospital, with nothing to do but follow the rehabilitation regime. When I woke up with detailed dream recall, I tape-recorded the report at once. That could be why it is about twice as long as most of my dream reports. Never in my life have I so much wanted to recall a dream, because never in my life has it meant so much to me to have one. I was getting better. My brain stem was healing.

In the Medieval Town dream, I could see, move, and feel everything intensely and realistically. As usual, the hallucinatory vividness completely fooled me: I thought I was awake and actually experiencing these improbable and awesome events. My delusional belief in the objective reality of my subjective experience is all the more surprising because, in my waking life, I am almost never unsure of what country I am in. In fact, it has happened to me only once. When I was watching the Italian film *Cinema Paradiso* in San Francisco, the film was so gripping that it made me believe I was in Italy. When I got up to go to the men's room, I was surprised to recognize that I was actually in California.

The point of this exception is that it proves the rule: The realistic but artificial scenarios of dreams and theater dictate conventions of orientation. The time is now, the place may be specified (accurately or inaccurately) or unspecified (with features that are incongruent with any real place), the characters are fixed (my wife and I) or fluid (the unidentified stranger), and they may behave toward one another in ways that defy reality. For example, in waking my wife is conspicuously faithful. It is I who is fearful of infidelity, but I ascribe this behavior to her. My emotions dominate this dream and give it its consistency and meaning, despite the psychotic components.

Some formal details that constitute classic dream bizarreness include the bridge that becomes a boat. This is a transmogrification in that both bridges and boats convey people over water, but one is fixed and the other is mobile. In real life, if a bridge suddenly floated away from its foundations, my attention would be drawn to this change, and I would be made very anxious by it. Drill bits are unlikely objects in a tourist setting, and they are not generally used to make holes in shoulder bags! I do collect tools, among them antique and modern drills and drill bits, but I don't take them with me when I travel (although I do wear a shoulder bag). But in the dream, it is the stranger who wants to seduce my wife who wears the shoulder bag. So the dream objects are unusual, confused, and scattered in a formally incongruous way.

Lia cooking in the kitchen of an inn is not just a flimsy excuse, as my dream interpretation suggests. She never cooks when we are on vacation, even in hotels where in-room cooking is possible. The architecture of the inn is also improbable: on one side we enter to find rooms including an unlikely kitchen; on the other side, I see a set of seats, adorned with flowers under a roof as in the Globe Theatre of Shakespeare's time but never seen today, except perhaps at tennis clubs.

Forgive me for belaboring the point about dream bizarreness, but content analysis, like all human efforts at understanding, tends either to make things look reasonable and coherent even when they cannot possibly be so or to interpret bizarre items as if they were symbolic transformations of unconscious wishes.

Lia and I do travel a lot, and within the past year we have visited Prague and southern Czechoslovakia (Bohemia), which is not unlike the scenario of the Medieval Town dream. In fact, it was in Bohemia that we had one of our most upsetting marital disagreements. But instead of being about infidelity, as that theme is played out in my dream, it was about how many children it made sense for us to have. For me, in my late 60s, the twins were more than enough. For Lia, in her early 40s, having more babies was still an attractive possibility.

The dream focuses on another marital issue: my own fear that my disability will make it impossible to hold onto Lia. This fear is both strong and conscious in my waking life. In the dream it is represented as Lia's vulnerability to seduction by another man. Despite the historical fact that I was a man who found her impossible to seduce, my dream fear and my own history of infidelity make her the one who might stray from the marital fold.

The 579 words of the report makes it highly likely that this was a REM sleep dream. Consistent with that assumption, the plot is complex, sustained, and richly detailed. The intensity of the hallucinatory imagery convinced me I was awake, and I never questioned that assumption despite abundant evidence that it could not possibly be true. My thought processes were markedly impaired: my self-reflective awareness, my judgment, and my logic were all poor, as is typical of REM sleep dreams. Moreover, the classic bizarreness of REM sleep dreams was present: the setting was unspecified (it could have been Hungary or Yugoslavia, but it wasn't definitely either), key personnel were unidentified and their behavior was odd (the man of whom I was jealous was a total stranger), the drill bit and the hole in my shoulder bag make only Freudian sense, my wife's cooking was unlikely in a hotel, and the architecture was incongruous. The consistent and strong emotional binding of these disparate elements is what gives the dream its obvious meaning: In my state of impaired health, I was worried about losing my most important companion and strongest supporter—my wife.

Scientists still quarrel about how to deal with dream content. The spectrum of confusion runs all the way from Freud-derived interpretative analysis of individual dreams to our own formal analysis of generic dreaming. One of the few measures that all scientists trust is word count. Word count is simply the number of words in a report. At 579 words, my Medieval Town dream is relatively long. Many reports have less than ten words. Does this low word count signal poor recall, incomplete awakening, or a low level of mental content? It is hard to know, but such short reports are often given in the garbled voice of someone who cannot wake up. This phenomenon needs further study to test the hypothesis that when sleep inertia is strong, awakening-based reports are worthless or even downright misleading. It is possible that subjects confabulate reports in order to be able to go back to sleep.

When we conducted our Nightcap-monitored study of home sleep and dreams, we performed word counts and found that report length increased with the time of night (the later, the longer) and with REM sleep (versus NREM sleep—REM reports tend to be longer, whatever the time of night). REM sleep reports are not only longer but also stronger (with respect to formal dream features such as hallucinosis, defective thinking, bizarreness, movement, and emotion). To those skeptics who say there is really no difference between late-night NREM and REM reports, we reply that we can accurately predict the sleep stage underlying a report from the report's formal features. We do make some errors, such as wrongly predicting that long, detailed NREM sleep reports come from REM. We don't yet know why some NREM reports are both long and formally REM-like.

A Canadian colleague of ours, Tore Nielsen, is investigating what he calls covert REM processes in NREM sleep. For example, we know that small twitches of the eyelids are visible in the Nightcap records 30 seconds before REM is evident in any of the usual polygraphic measures. A more important point is that all of sleep is a mix of NREM and REM physiology. Many eye movements occur in NREM sleep (making the term non-REM a misnomer), and NREM features, such as EEG spindles, may occur in REM sleep. The perceptive reader will recognize that this mixture of REM and NREM features is still another example of the tendency for large, complex brains such as ours to be in two states at once. Just as waking and sleep features may mix at sleep onset and offset, so may REM and NREM features mix as the sleep cycle evolves over the night.

It is this constant dynamic interplay between REM and NREM processes that gives sleep and dreaming such rich physiological possibilities. When my brain stem was first damaged, I could not sleep for ten nights. Thereafter, I could sleep fairly well. I suppose this means that my NREM sleep capacity had recovered. What I couldn't do between days 10 and 38 was oppose the NREM sleep process and create REM because my brain stem was damaged. It still is damaged, but it has recovered well enough to oppose NREM processes and let me dream again.

The same scientists who say that NREM and REM dream reports differ only in word count insist that NREM and REM sleep differ only in the intensity of activation. But neuromodulation also plays a role in differentiating the two cardinal aspects of sleep. Neuromodulation is the set of chemical processes that accompanies activation in REM. Aminergic neurones, which release norepinephrine and serotonin, are all but completely silent in REM, while cholinergic neurones become at least as active as they are in waking. Thus, while the cholinergic system obeys the activation rule, the aminergic neurones do not. This surely makes a difference. And it is an important difference for our dream theory.

In order for me to hallucinate the floating bridge, to travel on it, and to realize the strange building and the man I fear my wife may love, I need more than mere activation: I need to have enough internal visual stimulation to see clearly. This internal visual activation is actually stronger in REM than it is in waking. To experience apprehension and fear as keenly as I do, I must also have selective activation of my limbic brain. My failure to "wake up" and notice that I am dreaming, my ability to believe the internally generated scenario, and my inability to access either recent memory or critical thought are impossible to explain by the activation-only theory.

If activation were the only brain factor at work in determining dreaming, we would expect subjects to either wake up or have more awake-like mental experiences. Neither is the case. We don't wake up. Instead, we dream. This again suggests that there simply must be more to REM than mere activation.

To capture the brain basis of dreaming, I constructed a three-dimensional model. In this model, activation (A) is one dimension, acknowledging that turning up the brain's energy level turns up the mind's processing capability. The model also guarantees sleep in the presence of REM activation by changing access to and from the brain via input-output gating (I). Sensory stimuli are fenced out and movement is fenced in during REM. This puts the activated brain offline. The activated (A) offline (I) brain processes information differently because of changes in modulation (M) caused by the precipitous fall in aminergic neuromodulator and the reciprocal rise in cholinergic neuromodulation. In other words, we add to activation (A) factors (I) and (M) to account for the fact that we process information differently (M).

In the AIM model, time is the fourth dimension. Our nightly travels through AIM state space are seen as elliptical trajectories with progressively less deep incursions into the NREM domain and longer, deeper incursions into the REM domain. With only four dimensions, the model cannot deal with other important changes in physiology, such as the regionally selective activation in REM. Although we hope that this factor turns out to be controlled by fact M, it is already clear that an accurate model of brain-mind state space will be, like models of other complex systems, n-dimensional. Nevertheless, AIM is a strong step forward in the neuroscience revolution.

After recording the Medieval Town dream, on day 38 post-stroke, I was sleeping increasingly well—almost too well. At home from day 75 onward, I went to bed with my 5-year-old twins at about 9:00 or 10:00 p.m. and slept until 7:00 a.m. While I did have some fragmentary dream recall, I have not remembered another dream as well as this one. The possibility thus arises that I became hypersomnolent and was unable to recall my normal REM sleep dreams because I slept so deeply.

It is odd to be an experimental animal as well as the observer of a cruel experiment of nature. Of course, I am not the experimenter. No one is, but I now realize what my animal subjects went through in helping me understand how the brain stem controls sleep and dreaming. In a sense, these experimental animals helped me become one myself.

Now that we know what brain regions are activated and deactivated in the states of waking, sleeping, and dreaming, we need more detailed reports of mental activity following damage to the brain. All stroke patients whose language function is preserved can thus contribute importantly to our growing knowledge base.

Sad to say, my troubles were not over when I had recovered enough of my brain stem function to sleep and dream normally as well as to walk and talk relatively normally. Within the same year I experienced a combination of aspiration pneumonia and heart failure, which were almost fatal. The most likely precipitant of this medical disaster was the aspiration of food into my lungs. This is caused by the paralysis of the swallowing muscles of my throat (which are normally innervated by my damaged medullary brain stem). After five days in intensive care, and all manner of bizarre neuropsychiatric experiences, I emerged somewhat more incapacitated.

In addition to the ataxia caused by the stroke, I now have weakness owing to heart failure. Often it is all I can do to get myself upright and navigate my teetering gait. In addition, I have developed right facial pain called trigeminal neuralgia by the neurologists. This condition affects almost 25 percent of people with the symptoms of my kind of stroke, named Wallenburg's Syndrome after the kindly German doctor who first described it.

Before I published my paper, there was only one other first-person report of the subjective experience of Wallenburg stroke victims, and that one had nothing to say about effects on sleep and dreams. And none of my wonderful team of doctors can say why my heart is functioning so poorly. For want of a better theory, they say I had an infection of the heart muscle (myocarditis). I am skeptical of that diagnosis. Knowing that the brain stem controls blood pressure, heart rate, and rhythm via the same aminergic and cholinergic neurones that modulate the brain in sleep, I believe my heart might be misbehaving because of faulty instructions from command central—namely, my damaged medullary brain stem. We will probably never know who is right about this. An autopsy showing healthy heart muscle would favor my theory, but it would not prove the point.

Meanwhile, I am grateful to be able to read, think, and write with an ease that is very close to my previous level. And I love to sleep, which I do all too easily because I am taking a drug called Neurontin for my facial pain. It works. When I tried to get off it, I was too bothered by intrusive pain to concentrate on my work. So I'm back on the drug, and as I write, I am admirably resisting the temptation to join my 7-year-old twins in a nap. The Sicilian sun is blazing today, but there is a brisk northeast wind that has cleared the air of the Saharan haze and humidity that make 100 degrees feel like 120.

Nowadays, I remember very few dreams, probably because I sleep so deeply. But when I catch one, it's a scorcher indicating that my brain stem REM generator is doing just fine. I trust that the presentation and discussion of the Medieval Town dream will show that I regard dreams as emotionally salient and informative. The central meaning of my dream, fear of losing my wife, is as transparent as it is powerful. Of course, I have had the same fears in waking, so there is no dynamically unconscious meaning to be taken from this account.

When we, as medical students, were contemplating careers in psychiatry and neurology, we answered the question "What is the function of the heart?" with the joke answer "To pump blood to the brain." Now we can ask the converse question "What does the brain do for the heart?" and answer "Command it to pump blood to the brain." The brain and the heart have one important element in common: pacemaker cells. The Purkinje cells of the heart and the aminergic cells of the brain therefore depolarize without external stimulation. By firing spontaneously, these cells establish the autonomous rhythmicity of both the heart and the brain. Normally these two interdependent organs work well together.

Chapter 13

French Kiss
The Erotic Hypothalamus

*On Saturday morning, after Lia got up to cook breakfast,
I had two incredible dreams, in which I was kissing. In
both, my female collaborator was unseen and, in fact, disembodied! I could see only a mouth, wide open in a most
lascivious fashion. I was amazed to find that I could
induce, in myself, the most vivid and sensual feelings—
more intense than I have felt in years. In the second dream,
which followed a short wake-up interval, the intensity was
quasi-orgasmic, although I did not have an orgasm.
During the second kiss, when I beheld the woman for a
fleeting instant before the act, I remember thinking it was
surprising that a kiss, however French, could be so erotic.
And so I tried rimming the open oval lips with my tongue
and found, to my surprise, that I did not even have to
touch the lips to feel electrifying sexual energy throughout
my body! It seemed impossible, but, despite the fact that
this was the second time around, and I must have been
very lightly asleep, I never dreamed I was dreaming.*

I had just returned from Lyon, France, where I participated in the international meeting called The Paradox of Sleep in honor of Michel Jouvet, who was retiring. After the meeting, I spent two days with my old friend and colleague, Francois Michel, often in the company of his companion, Marie-Anne Henaff.

The visit, and the meeting, resuscitated many strong, emotionally salient memories, chief among them my first extramarital affair with a French woman in 1963. The erotic aspect of this affair was intense, even searing. It was my first sustained erotic liaison. It lasted six years, until my lover's untimely death in 1969.

Having had a happy second marriage and, since my stroke in February 2001, a markedly lessened libido, I had no intention of undertaking a new seduction or even rekindling an old one. So I was surprised to perceive that Marie-Anne was affectionate in a very inviting and solicitous way. I did not reciprocate and still was not convinced of her intentions until we reached the airport on the day of my departure.

There, after a flight delay and a stop at a café, we said goodbye. The goodbye kiss was neither perfunctory nor chaste. Yet it was I who withdrew. On the plane I asked myself if I had been, somehow, misleading. Had I suggested an interest in physical intimacy when all I really wanted was a warm friendship?

Once home, I received a love letter. I put off replying for two days. The solution that finally presented itself to me was a poem about a jasmine vine, which is using an old shrub in my garden to spread itself out in a luxurious blanket of white bloom. Two days later, I had the French Kiss dreams.

This dream is atypical in several important respects, but it shows how powerful, emotionally salient, and unexpectedly instinctual dreams can be. They are brief, lasting, at most, an estimated 2 minutes each. They are dissociated from the usual plot features of REM sleep dreams. There are no characters except for the fleeting appearance of an unidentified female in dream 2. The mouth that I kiss so erotically is, in fact, totally disembodied in both instances. There is only the mouth! No location, no scenario, no time, no people! It is a dream of almost pure somatic feeling.

The sensations, as I perceive them, begin with an innocent kiss in which I feel affection but no bodily pleasure. Then, especially in the second dream, I am aware that if I want the kiss to turn me on, it will. And it does! The sense of erotic excitement spreads like wildfire through my entire body.

Only rarely does a dream progress to orgasm in my experience. The erotic brain—let's guess that sex is mediated by the hypothalamic region—can be activated in sleep. Furthermore, let's suppose activation can dominate dream plot construction. The French Kiss dream seems to say that, moreover, an incidental encounter with a former friend who still harbors romantic feelings can trigger a remarkably different sort of erotic dream, even in an old man with a large hole in his head—that is, in his medulla oblongata.

The erotic feelings engendered by my two French Kiss dreams were every bit as intense as an experience I had at the Casino de Paris when I was 18. I was seated far from the stage in the second balcony, but the sight of the naked showgirls was enough to stimulate full sexual release without my touching my body. I mention this experience because it was the only one in my lifetime when my waking erotic experience approached that of my dreams. And I mention it now because, 50 years later, I can still experience erotic satisfaction in my brain alone.

This tells us something about the erotic brain. Is this the disguise and censorship of forbidden desire? Not at all. It is the revelation of rekindled desire that never quits, even though the conscious mind and the body are not up for it! My conscious decision not to reciprocate romantic overtures counts for little.

This realization confirms the power of instinct over volition, which we must credit Freud for recognizing and emphasizing. But the dreams indicate that much of desire, erotic excitement, and quasi-orgiastic ecstasy takes place in the brain-mind purely as a function of regional brain activation. The upshot is that the study of dreams is the study of instinctual life as well as the study of hallucinations, delusions, and diminished thought.

Despite the fact that both of these red-hot dreamlets occurred in the early morning after I had already awakened at least once, the hallucinated kiss was so convincingly real that it never occurred to me to doubt it, or to see it as a response to a recent overture by an old friend, or really to question it at all. It was too good for that. And I am glad that I had these dreams, not Freud!

The 1950s popular song "A Kiss to Build a Dream On" emphasizes the power of a kiss to trigger a dream of love that long outlasts the kiss. In my dream, it is the memory of an old love that triggers the hallucinated kiss. No doubt it was the relatively chaste and virginal ethic of the post-war era that made such a song so appealing—and so indelible in my mind that it came into my head one morning recently when I was making the bed in which only a few days before I had had the electrifying kiss dreams. The song says it all. With just one kiss comes the hint of openness to more serious and complete lovemaking in the future—or in the virtual reality of imagination.

My kiss dream, like the song, shows how remarkable imagination can be when it builds a dream on "just one kiss." I remain uncertain as to why Marie-Anne gave me that kiss. But I know it meant to me that I am still capable of passion—even with my damaged brain-mind. That's reassuring in itself, but it is also evidence of how resilient the yearning brain-mind really is. This dream has strong implications for recovery from explicit and severe injuries to the brain (like my stroke) and for coping with the more-subtle dysfunctions of old age, such as impotence (which I also have). Since all our experience really occurs in our heads, experience that occurs only in our heads can be as real as that which is tied to external reality. We need to learn more about how to access our native capacity to create a virtual reality that is every bit as enjoyable as the real thing.

The French Kiss dream also reemphasizes the main point of the difference between our new brain-based model of the mind and its psychodynamic antecedents. The essence of the difference is the distinction between reflex and spontaneous activity. If the French Kiss dreams were triggered by the real airport kiss in France, they obey reflex laws but with a much longer delay than reflex theory would normally allow.

We must assume, therefore, that the spontaneous occurrence of erotic feelings in sleep builds on the memory of experience in a strange and instructive way. Instead of a day residue pairing up with an unconscious wish, we suggest that erotic feelings, arising spontaneously, pair up with recent memories to create dream eroticism.

This example further increases our conviction that Walter Cannon was correct in assuming that all emotion, including erotic emotion, is experienced centrally, not in the body as our subjective experience in waking and William James's theory would have it. In my dream, I experience erotic emotion that arises entirely within my own brain. My body is cut off from my brain by sensory inhibition.

Far from being a turnoff or a deterrent to sexual energy, as Freud told us, dreaming can be a booster for sexual sparks, fanning them into fires. So why aren't more dreams sexual? We ourselves were surprised to learn, in our study of dream emotion, that remembered dreams were erotic in only 5 percent of the cases! But how much of waking consciousness is occupied by erotic fantasy or erotic behavior? Obviously, we would expect large individual differences in these statistics. But on the whole, sexual feelings and fantasies are surprisingly rare and fleeting. Our subjects were not hormone-soaked adolescents, but they were not old dodders either. They were adults in their sexual prime, all of whom hoped—and loudly asserted—that sexual dreams were far more common than their own data showed them to be.

The fact is that there is a great deal of time—during the night and during the day—when the brain-mind occupies itself—quite adaptively, thank you—with important tasks other than sex. These tasks include such basic mental housekeeping functions as review and rehearsal of social interactions, planning for upcoming work, and staying on track toward other goals that are every bit as important to survival and procreation as sex itself.

Surely some societies are repressive. Freud's Vienna in the late 1890s may have been one. But there are good reasons to doubt even that assertion. Consider the hypersexuality of Vienna writers such as Arthur Schnitzler and artists such as Max Klinger and Alfred Kubin. Now we live in a society that did everything in its power to liberate sexuality, especially in the 1960s, but still there are limits—innate, natural limits—to how much sexual fantasy or activity can be maintained. It's as disappointing as it is liberating to recognize those limits. And then to see that even when sexual life appears to be over, just one kiss is enough to build a dream on! In my case, not just one dream, but two of the most erotic dreams I have ever had.

Recent memory may provide seeds for this process, but the dreaming brain-mind does not simply replay previous experience in anything like mnemonic form. Instead, it elaborates, associates, and—here's the key word—creates a world of its own devising. That this autocreative process is lifelong, universal, and very probably formative of us as sentient people means that Shakespeare was right when he said that "we are such stuff as dreams are made on."

The functional autocreativity of the brain-mind provides evidence that healing of real and imagined wounds can occur. Even if the brain is fixed and limited in its structural nature, the diversity and plasticity of its states give it a power possessed by no other organ in the body. Imagination, which is devalued by some because it is virtual, is in fact the highest of all human talents. Life is short, but art is long. When disease strikes, the brain reorganizes and, like dreaming itself, makes the best of a bad job.

There has been a recent flurry of hope that brain cells might after all have the possibility of regeneration. But even if this is so—and even if stem cell science proceeds to reseeding the brain as if it were a lawn—it is hard to imagine how such processes could be selective and targeted to the specific brain regions needing repair (as in my stroke). While hoping that I am wrong to be so pessimistic, and while waiting for my brain transplant to arrive by helicopter, I will sing another pop tune from the '40s: "I Can Dream, Can't I?"

Freud Reawakens

If I were alive today, I would be 148 years old. I rejected the idea of an afterlife as the gratuitous wish of mortals. I thought that when I died I would be dead and gone forever—except, of course, for my scientific work. Indeed, my speculative theory of mind has had outstanding longevity. Perhaps undue.

I often argued that the problems of creating a truly scientific psychology would be solved only by knowing enough about the brain's physiology and chemistry. I even predicted that all my psychoanalytic ideas would one day be replaced by the formulation of physiology and biochemistry. So many things were learned in the 20th century that were directly relevant to the task I first set for myself in 1895 when I was 39. At the time I published *The Interpretation of Dreams*, in 1900, I was 44 and convinced that psychoanalysis was the method of choice for work on a science of the mind. Although I am sad to see how mistaken I was in constructing my dream theory, and the whole idea that human experience could be explained in terms of wish fulfillment, I am pleased to see that my Project for a Scientific Psychology is now moving forward at such a thrilling speed.

Moruzzi and Magoun's discovery of the reticular activating system and Aserinsky and Kleitman's discovery of REM sleep made scales fall from my eyes. If the brain was periodically self-activated in sleep, that meant that dreaming could occur at intervals throughout the night and not just at the time of awakening as I had assumed.

Of course, the brain's electrical activity was discovered during my lifetime. But I tended to ignore Caton's 1875 discoveries, as almost everyone else did. How could we be sure that voltage fluctuations recorded in the brains of rabbits had anything to do with the human mind or anything else of interest to us psychiatrists? In retrospect, anyone can see that those fussy Englishmen were on to something.

A watershed came in 1928 when my fellow psychiatrist, Adolf Berger, described the human EEG. Berger worked in Jena, which is not far from Vienna. Berger convinced his skeptical critics that the voltage changes he observed when he recorded from electrodes attached to the scalp were real and of brain origin by showing that the brain waves always slowed and grew larger when his subjects fell asleep. But by then I was organizing an international movement and applying my own psychoanalytic theory to the social sciences and even to religion. In fact, while Berger was describing brain changes in sleep—which certainly would have caught my attention in 1895—by 1928 I had already published my psychoanalytic deconstruction of religious belief as the delusional wish of people for an all-knowing, all-caring father.

Another important point is that I had turned away from my scientific base in neurology—or tried to—because so many importunate colleagues wanted me to neurologize my theory. I fiercely opposed these offerings as unwanted and undesirable because they constituted a dangerous dilution or even destruction of psychoanalysis. I don't recall whether I even read Berger's papers when they came out, but in any case, they could hardly have influenced my dream theory, which I declared alive and perfectly well in my famous revision article of 1933. As some have since pointed out, that article contained no revisions. My only excuse, and it's a good one, is that the storm clouds were growing over Europe, I was 77, and it was critical for me to move to London in the interest of survival. Other matters were on my mind.

In my later years, when I was beset with cancer of the jaw, it was all I could do to keep myself alive and keep up what became the routine duties of leading the psychoanalytic movement. I had to be constantly on guard lest one of my collaborators contaminate the theory. At the same time, I was eager to bring new people into the movement, even if they were not physicians or even psychologists. I never even noticed that the American scientists Loomis and Harvey described a change in brain wave activity during the night, which was periodic at about 90–100 minutes. Their important and premonitory paper was published in 1936, a few years before I died.

Here, finally, was the way to prove psychoanalytic dream theory, not discredit it. When the results of the animal work began to come in during the late '50s and early '60s, I would have had a hard time remaining complacent, because several key aspects of my theory were challenged by the findings.

The first, of course, was the wish-fulfillment idea. This was the basic motive of mental life for psychoanalysis: We want things we can't have, the desire is thwarted, and the energy goes underground and stays there—volcano-like—until an opening is provided. Then, boom! Out it pops. One of those openings is sleep. Although it is still possible to assert that forbidden desires often find their way into dreams, it is much more difficult to maintain that they actually cause any of them. If that were the case, why would REM sleep dreaming be periodic? Why would it be more prevalent in babies, and even more so in fetuses? And why would REM be shared by all mammals?

In order to maintain that unconscious wishes cause dreams, I would be forced to maintain that the brain (in REM) and the mind (in dreaming) were completely dissociated from each other. This is a position taken by Mark Solms, who is doing his best to defend my ideas (as well as to manage my library in London). But can unconscious wishes and dopamine be equated?

As for my idea that the function of dreaming was to protect sleep (my "guardian of sleep" notion), it was soon apparent from the work of Michel Jouvet that the brain has its own clever ways of maintaining sleep while simultaneously turning itself on and off. These mechanisms of REM sleep generation were embedded in two processes that interested me immensely. The blockade of motor output, which Jouvet described at the physiological level, could mediate the inexpressibility of motor actions, which I postulated as an important psychological reason for disguise and censorship of the dream-instigating wishes. It is certainly true, as Solms points out, that motivated behaviors depend on dopamine. And dopamine output does not decline in sleep the way norepinephrine and serotonin do. So maybe motivation is enhanced—or unmasked in REM—but that's a far cry from the instinctual wishes I had in mind.

In order to explain the intense imagery of dreams, I proposed that the mind regressed defensively to the point of visual hallucinosis. It now seems that what was really going on was a shift in brain state that made visual imagery primary and positive. In other words, the brain was actively self-stimulating as an integral part of the REM sleep-generation process. The sensations of dreams are the direct result of brain activation, not a secondary psychological response.

Many of my defenders have said that REM and dreaming are dissociable. It is true that dreaming does occur at sleep onset and in late-night non-REM sleep. But if I were alive now and wanted to help my patients remember dreams, I'd have them wake up in REM. The statistics indicate that this is the best bet, and the scientist in me says that the chemical changes in the brain that lead to REM sleep dreaming are 50 percent as strong in NREM as they are in REM.

Some object to any kind of physiological rendering of what they take to be irreducibly psychological processes. For them, brain research has never had anything to offer psychoanalysis and never will have anything to offer it. Their disdain for physiology leads them to an erroneous rejection of sleep science because they consider it reductionistic. There are two reasons why I would now dismiss this objection.

One is my recognition that reductionism is the very soul of science. When I formulated my dream theory, I wanted, as all good scientists do, to explain as many variables as I could with the least possible number of assumptions. My dream theory was certainly simple. And the central notion of the theory—namely, disguise and censorship—could not only be used to explain dreaming but also could be extended to all the vicissitudes of mental life.

As I have often insisted, I would have used sleep neurophysiology in my theorizing if I could have. But basic neurophysiology and all the brain imaging and electrical monitoring techniques of the 21st century were not available in 1895. It is now clear that a sea change has occurred in neurobiology. The automatic brain activation in sleep helps explain the occurrence of dreaming. Wishes still might play a role in shaping dream content, but they can no longer be considered initiative. The brain turns itself on and off all by itself. So my theory about dream generation was incorrect.

And the brain also both regulates the internal perceptions causing visual hallucinosis (which I regarded as regressive defenses) and inhibits motor output (which I ascribed to censorship). So the net effect, which I am proud to have appreciated even in 1895, is that the dreaming brain-mind is consciousness offline, as it were. Now I see why. Its inputs and outputs are gated by reliable, automatic physiological processes.

Cellular and molecular work that began in the late 1960s gave rise in the 1970s to well-articulated objections to my central hypotheses. I myself was a cellular neurophysiologist. Many people know that I was a neurologist. I became a psychiatrist only because so many of my so-called neurological patients were afflicted by functional disabilities that we could not then explain neurologically. But few people realize that I, like many 20th century psychiatrist-neurophysiologists, thought that understanding the brain at the cellular level was necessary to developing a brain basis for psychology.

In the late 19th century the scientific choices were limited to invertebrate preparations like the crayfish I worked on in Brucke's lab in the physiology department of the medical school in Vienna. Even today, I notice that some psychiatrists like the Nobel laureate Eric Kandel have chosen to work on subhuman, submammalian creatures. My own research was frustrating not only because my career could not advance in the physiology department but also because the neurophysiology of the crab's claw-opening muscle was insufficiently enlightening to form a brain-based theory of the mind. When I began to study dreams, I was as struck by their apparent strangeness, as was everyone else who had even written about them. I was interested in my own dreams. But I did not carefully collect, catalog, or count dreams or dream characteristics. Instead, I viewed dreams as stories and assumed that the peculiar surface structure was the result of a deep transformation from unacceptable unconscious wishes to apparently nonsensical conscious experience.

The mind seemed to be playing tricks on itself. First, it censored and then disguised the unconscious wishes that I took to be the dream instigators. I am pleased that some modern neuropsychologists, led by Mark Solms, are still championing the wish-instigation idea, but if disguise-censorship is dead, it hardly matters whether dreams are motivated by wishes. As I should have more forcefully argued in 1933, some dreams clearly are driven by what can reasonably be called wishes, but many are not. The prominence of negative emotions such as fear, anxiety, and anger cannot be explained by my theory.

For now, let us consider what I take to be my greatest intellectual error, an error so great that a century of potential progress was held back. This was the famous secret of dreams that came to me (like a religious revelation) on Pentecostal Sunday in 1896. I suddenly became convinced that the nonsense of dreams was only apparent and that what I called the manifest content and purposefully misleading was a pale copy of the latent and unacceptably true content of the dream wish.

Why I ever thought such a thing is beyond me now. I should have realized that this idea was itself a delusion. No wonder they gave me the Goethe Prize (for literature) and not the Nobel (for science). I am embarrassed to admit that I completely ignored the advice of the American philosopher and psychologist William James, whom I met in Worcester in 1909. James, in his own comical way, said that my notion of a dynamically repressed unconscious at the center of mental life was "a tumbling ground for whimsy." In my own defense, the distinction I drew between the manifest and latent content of dreams was based on the conviction that every aspect of mental life was a compromise between our deepest drives and social convention. I still hold this proposition to be true. But I must admit that I opened the door to wild and unsupported speculation—anything but the solid science I hoped to build!

To a neurologist, the possibility that it was the physiological changes in the brain during sleep that changed mental activity in dreaming is much more interesting than I then allowed. I knew that dreaming was a state akin to madness. It was therefore subject to analysis using my own traditional tools, including the mental status exam—a fundamental part of neurology from the very beginning.

If I had had my wits about me, I would have asked myself and my colleagues if the bizarreness of dreams might not spring from organic changes in brain activity. In 1900, I could not guess what was going on in the brain, such as the aminergic demodulation that was demonstrated in Boston 75 years later. Specific experiments were needed to reveal the dramatic changes in brain chemistry implied by the REM sleep-related arrest of firing of serotonergic and noradrenergic neurones. As researchers came to note, an aminergically demodulated brain is likely, if otherwise activated, to have its own distinctive mental accompaniment.

If I had peeked into the crib of my own children, I could have seen REM for myself. And when Carl Jung and I later both realized that dreaming was some kind of psychosis, we could have asked ourselves what kind of psychosis it was. If instead of jumping to my sweeping conclusion that the bizarreness of dreams reflects the operation of defensive disguises via displacement, condensation, and symbolization I had just looked more carefully at dreams, I would have seen that bizarreness reduced to disorientation! As a neurologist, I knew that the cause of disorientation was the loss of recent memory. And that signaled organic mental illness.

The time has come to clear the decks of the wreckage of psychoanalysis and build a new science of dreams based on what is now known about the brain. After all, this was my own starting point, and any effort of this kind must recognize how bold and astute I was to try. Let me outline the task of revision and comment on how I see it going.

My attribution of dream forgetting to defensive repression now seems ridiculous on two counts. In the first place, if the dreams had already been bowdlerized by disguise and censorship, why would people need to repress them? By the very nature of the dream work, these toothless bits of nonsense might just as well be jettisoned. And what better way to jettison them than simple amnesia? Even more embarrassing to me is the recognition of the recent memory deficit that operates within dreams and for dreams post-awakening. This is a dead giveaway that something powerful is going on in the organic domain. Carl Jung and I both knew that organic psychoses, like those associated with alcoholism and drug abuse, were characterized by recent memory defects, disorientation, visual hallucinosis, and confabulation. We should have realized that dreaming was more like that kind of delirium than the so-called functional psychoses, the schizophrenias that Jung had studied with Bleuler, and the mood disorders that attracted us as grounds for application of our psychoanalytic theories.

No one else got it exactly right either. The long introductory chapter in *The Interpretation of Dreams* has two striking ambitions. One is to discredit all my predecessors in the domain of dream theory, and the other is to dismiss many of those predecessors who were neurologically oriented. One particularly virulent attack of mine was aimed at Wilhelm Wundt, who is now enjoying a posthumous resuscitation. I needed to discredit Wundt because he came so close to getting it right using physiological arguments.

In other words, I was determined to fault any and all neurological theories since I myself had been unable to use brain science to build a bottom-up formulation of the mind. With all the discoveries that have been made in the fifty years between 1954 and 2004, it is now possible to build a new dynamic psychology on the solid base of brain science. I was, may I say, ahead of my time.

Neurology has become dynamic and functional rather than static and structural. Let me explain what I mean by this. At the end of the 19th century, and even well into the middle of the 20th, the only pathology recognized by neurologists was that caused by structural damage to the brain. We could see the effects of strokes, infection, and tumors on the brain. Because we couldn't see anything of the structured kind in hysterical paralysis, it was understandable that we tried to develop what we called psychodynamic models to explain such problems. We never took into account the dynamics of the brain itself because we didn't recognize that altered states of mind implied altered states of the brain.

Hysterical paralysis and other losses of function were not considered organic because we didn't know that the brain, by its very nature, was prone to dissociation, paralysis, and memory loss. The proof of this fundamental fact is given by the physiology of sleep and, more significantly, by the physiology of REM, in which functional paralysis and memory loss are normal (not to mention the visual hallucinations and delusions). Instead of assuming that every state of the mind depended on a state of the brain, I tended to assume that the mind was independent of the brain-mind and could be investigated on its own.

That is the trap of the dualism invented by Descartes. He believed that the brain and the mind, while not interactive or identical, were perfectly synchronized. His dualism was just as much anathema to me as religion, and I tried hard to avoid both. But even a genius can be misled when he tackles something as difficult as the brain-mind problem. Because neurobiology had let me down, I made the mistake of turning my back on it. I have been reading *13 Dreams Freud Never Had*. The really surprising and satisfying thing about reading Hobson's accounts of his own dreams is his distinction between the formal aspects ascribed to alterations in brain physiology and the individually meaningful dream content. I must confess I was amazed to see how much psychological insight could be garnered from what I had dismissed as merely manifest content. In retrospect, this is really all I did in interpreting my Irma dream. I was anxious about a clinical error of judgment, and I dreamed about it. No symbols. No unconscious wishes. No real interpretation in the psychoanalytic sense. There might be more going on beneath the surface, but why take a chance, especially if that leads us to walk right by the treasure trove of dream meaning sitting there on the surface?

The key is the novel recognition of the salience of emotion in dream content formation. This resonates with my insistence on the primacy of affect (I called it instinct) in shaping dreaming (and waking thought). I myself had trouble with the presence of so much negative affect in dreams. It obviously didn't fit with my wish-fulfillment hypothesis, but now I see that activation of the limbic system is as likely to unleash anxiety and anger as it is to promote joy and elation. Both are important built-in systems that help as well as hinder us.

The fact that activation synthesis cannot yet say why sometimes negative and sometimes positive emotion is triggered hardly matters. Given the rapid expansion of brain-imaging technology, my guess is that an explanation is forthcoming.

By embracing dynamic neurology (instead of trying to fit the new data to my antique and outmoded theory), we can have it all: neurology, a dynamic model of the mind, and a way of helping people find meaning in their lives. All of this is now possible while avoiding the embarrassment of my understandable but regrettable exaggeration of sexual impulses as a major drive of everything. I can claim that it is largely because of me that the duplicity of Victorian mores regarding sex is out in the open. And I am proud to have sensitized whole generations of people to the still-powerful drive of eros.

Endnotes

Prologue

1. Freud, S. Project for a Scientific Psychology. In: *The Origins of Psychoanalysis: Letters to Wilhelm Fliess, Drafts and Notes: 1887–1902*, eds. Marie Bonaparte, Anna Freud, and Ernest Kris. New York: Basic Books, 1954, pp. 347–445.

2. Freud, S. *The Interpretation of Dreams*. London: G. Allen & Unwin, Ltd., New York: Macmillan, 1913.

3. McCarley, R. W., and J. A. Hobson (1977). The neurobiological origins of psychoanalytic theory. American Journal of Psychiatry, 134(11):1211–1221.

4. Freud, S. *On Dreams*. New York: Norton, 1963.

Chapter 1, "Dreamstage with Elephants"

1. Dreamstage Dream, JAH Journal Vol. 13, 4/3/1982.

2. Dreamstage. An Experimental Portrait of the Sleeping Brain. Scientific Catalog by J. Allan Hobson, Paul Earls, and Theodore Spagna. © J. Allan Hobson and Hoffman La Roche, Inc., 1977.

3. Dreamstage: Exhibit Catalog by J. Allan Hobson, Paul Earls, and Theodore Spagna. © J. Allan Hobson and Hoffman La Roche, Inc., 1978.

4. Hobson, J. A. Dreamstage or Letting the Brain Speak for Itself. Unpublished manuscript, available on request.

5. Dreamscreen: Le Reveur de Lainé, Exhibit Catalog by J. Allan Hobson, Paul Coupille, Ragnhild Reingardt-Karlstrom, Ted Spagna, and Jean Didier Vincent. © Sigma, Bordeaux, France, 1982.

6.   Jouvet, M. Recherches sur les structures nerveuses et les mecanismes responsables des differentes phases du sommeil physiologique. Archives Italiennes de Biologie, 1962, 100:125–206.

7.   Sherin, J. E., J. K. Elmquist, F. Torrealba, and C. B. Saper (1998). Innervation of histaminergic tuberomammillary neurons by GABAergic and galaninergic neurons in the ventrolateral preoptic nucleus of the rat. Journal of Neuroscience, 18(12):470521.

8.   Hobson, J. A., T. Spagna, and R. Malenka (1978). Ethology of sleep studied with time-lapse photography: Postural immobility and sleep-cycle phase in humans. Science, 201:1251–1253.

9.   Hobson, J. A., R. W. McCarley, and P. W. Wyzinski (1975). Sleep cycle oscillation: reciprocal discharge by two brain stem neuronal groups. Science, 189:55–58.

10.   McCarley, R. W. and J. A. Hobson (1975). Neuronal excitability modulation over the sleep cycle: a structural and mathematical model. Science, 189:58–60.

11.   McCarley, R. W. and J. A. Hobson (1977). The neurobiological origins of psychoanalytic theory. American Journal of Psychiatry, 134(11):1211–1221.

12.   Hobson, J. A. and R. W. McCarley (1977). The brain as a dream state generator: an activation-synthesis hypothesis of the dream process. American Journal of Psychiatry, 134(12):1335–1348.

13.   Winfree, Carey. Starring in Show is a Real Dream Job. New York Times, May 16, 1977.

Chapter 2, "HMS Marathon"

1.   HMS Marathon Dream, JAH Journal Vol. 18, 11/13/1984.

2.   Of Flame and Clay—Dialogues on Mind-Body Interaction, Annals of the William James Seminar. Jeffrey Saver and Stephen Denlinger, William James Printers, Boston, 1986, Vol. 2, 1982–1986.

3. Saver, J. and S. Denlinger. Which Doctor Is Not a Witch Doctor? Advances, 1985, 2:20–30

4. Hobson, J. A. (1988). *The Dreaming Brain*. New York: Basic Books.

5. Hartley, D. *Observations on Man, His Frame, His Duty, and His Expectations*. London: Johnson, 1802.

6. Constantinidis, C., G. V. Williams, and P. S. Goldman-Rakic. A role for inhibition in shaping the temporal flow of information in prefrontal cortex. Nature Neuroscience, 2002 Feb; 5(2):175–80.

7. Pittenger, C. and E. Kandel. A genetic switch for long-term memory. Comptes Rendues dell'Academie des Sciences, 1998 Feb–Mar; 321(2–3):91–6.

8. Maquet, P., J. M. Peters, J. Aerts, G. DelFiore, C. Degueldre, A. Luxen, and G. Franck. Functional neuroanatomy of human rapid-eye movement sleep and dreaming. Nature, 1996, 383(6596):163–66.

9. Braun, A. R., T. J. Balkin, N. J. Wesenten, R. E. Carson, M. Varga, P. Baldwin, S. Selbie, G. Belenky, and P. Herscovitch. Regional cerebral blood flow throughout the sleep-wake cycle. An H2(15)O PET study. Brain, 1997 Jul; 120 (Pt 7):1173–97.

10. Nofzinger, E. A., M. A. Mintun, M. Wiseman, D. J. Kupfer, and R. Y. Moore. Forebrain activation in REM sleep: an FDG PET study. Brain Research, 1997 Oct 3; 770(1–2):192–201.

11. Winson, J. *Brain and Psyche: The Biology of the Unconscious*. Garden City, N.Y.: Anchor Press/Doubleday, 1985.

12. Jung, M. W., S. I. Wiener, and B. L. McNaughton. Comparison of spatial firing characteristics of units in dorsal and ventral hippocampus of the rat. Journal of Neuroscience, 1994 Dec; 14(12):7347–56.

13. Stickgold, R., J. A. Hobson, R. Fosse, and M. Fosse (2001). Sleep, learning and dreams: Off-line memory reprocessing. Science, 294, 1052–1057.

14. Darwin, C. *The Expression of the Emotions in Man and Animals*. London: J. Murray, 1872.

15. Jouvet, M. Essai sur le rêver. Archives Italiennes de Biologie, 1973, 111:564–76.

16. Revonsuo, A. The reinterpretation of dreams: an evolutionary hypothesis of the function of dreaming. Behavioral and Brain Sciences, 2000 Dec; 23(6):877–901.

Chapter 3, "Lobster Brain"

1.   Lobster Brain Dream, JAH Journal Vol. 19, 6/22/1985.

2.   Mukhametov, L. M., A. I. Oleksenko, and I. G. Poliakova. Quantitative characteristics of the electrocorticographic sleep stages in bottle-nosed dolphins. Neirofiziologiia, 1988; 20(4):532–8.

3.   Darwin, C. *On the Origin of Species*. Washington Square, N.Y.: New York University Press, 1988.

4.   Edelman, Gerald M. *Bright Air, Brilliant Fire: On the Matter of the Mind*. New York: Basic Books, 1992.

5.   Damasio, A. *The Feeling of What Happens*. New York: Harcourt Brace, 1999.

6.   Tononi, G. and G. M. Edelman. Consciousness and complexity. Science, 1998 Dec 4; 282(5395):1846–51.

7.   Pinker, S. *The Language Instinct: How the Mind Creates Language*. New York: Perennial, 2000.

8.   Mamelak, A. and J. A. Hobson (1989). Nightcap: A home-based sleep monitoring system. Sleep, 12:157–166.

9.   Flanagan, Owen J. *Dreaming Souls: Sleep, Dreams, and the Evolution of the Conscious Mind*. New York: Oxford University Press, 2000.

10. Gould, S. J. The exaptive excellence of spandrels as a term and prototype. Proceedings of the National Academy of Sciences, 1997 September 30; 94(20): 10750–10755.

11. Wilson, V. J., M. Kato, and B. W. Peterson. Convergence of inputs on Deiters neurones. Nature, 1966 Sep 24; 211(56):1409–11.

12. Lydic, R., R. W. McCarley, and J. A. Hobson (1987). Serotonic neurons and sleep. I. Long-term recordings of dorsal raphé discharge frequency and PGO waves. Archives Italiennes de Biologie, 125:317–343.

13. Wyzinsky, P. W., R. W. McCarley, and J. A. Hobson (1978). Discharge properties of pontine reticulospinal neurons during sleep-waking cycle. Journal of Neurophysiology, 41(3):821–834.

14. Kruger L. and T. A. Woolsey. Rafael Lorente de Nò: 1902–1990. Journal of Comparative Neurology, 1990 Oct 1; 300(1):1–4.

15. Hobson, J. A., R. Lydic, and H. A. Baghdoyan (1986). Evolving concepts of sleep cycle generation: From brain centers to neuronal populations. Behavioral and Brain Sciences, 9:371–448.

16. Livingstone, M. S., S. F. Schaeffer, and E. A. Kravitz. Biochemistry and ultrastructure of serotonergic nerve endings in the lobster: serotonin and octopamine are contained in different nerve endings. Journal of Neurobiology, 1981 Jan; 12(1):27–54.

Chapter 4, "Caravaggio"

1. Caravaggio Dream, JAH Journal Vol. 23, 6/27/1985.

2. Fosse, R., R. Stickgold, and J. A. Hobson (2001). Brain-mind states: Reciprocal variation in thoughts and hallucinations. Psychological Science, 12, 30–36

3. Fosse, R., R. Stickgold, et al. (2002). Emotional experience during rapid eye movement sleep in narcolepsy. Sleep, 25: 724–732.

4. Porte, H. and J. A. Hobson (1996). Physical motion in dreams: One measure of three theories. Journal of Abnormal Psychology, 105:329–335.

5. Dream Journal. L. L. Buchanan, Summer 1939 (original manuscript). Copies available on request.

6. Evarts, E. V. Activity of neurons in visual cortex of the cat during sleep with low voltage fast EEG activity. Journal of Neurophysiology, 1962, 25:812–816.

7. Dryden, John. *The Fables of John Dryden*. London: Printed by T. Bensley for J. Edwards and E. Harding, c1797.

8.   Jung, C. G. *Memories, Dreams, Reflections*. New York: Pantheon Books, 1963.

9.   Sulloway, F. J. *Freud, Biologist of the Mind: Beyond the Psychoanalytic Legend*. New York: Basic Books, 1979.

10. Merritt, J. M., R. A. Stickgold, E. F. Pace-Schott, J. Williams, and J. A. Hobson (1994). Emotion profiles in the dreams of young adult men and women. Consciousness and Cognition, 3:46–60.

11. Edelman, R. The Clinician's Guide to the Theory and Practice on NMR Scanning. Discussions in Neurosciences, 1984, 1(1):48.

12. Mazziota, J. C. PET Scanning: Principles and Applications. Discussions in Neurosciences, 1985, 2(1):47.

13. James, William. The varieties of religious experience; a study in human nature; being the Gifford lectures on natural religion delivered at Edinburgh in 1901–1902, by William James. New York: Longmans, Green, 1902.

Chapter 5, "Italian Romance"

1.   Italian Romance Dream, JAH Journal Vol. 23, 11/12/1985.

2.   Hobson, J. A. *Dreaming as Delirium*. Cambridge: The MIT Press, 1999.

3.   Merritt, J. M., R. A. Stickgold, E. F. Pace-Schott, J. Williams, and J. A. Hobson (1994). Emotion profiles in the dreams of young adult men and women. Consciousness and Cognition, 3:46–60.

4.   Solms, M. *The Neuropsychology of Dreams*. New York: Lawrence Erlbaum Associates, Inc., 1997.

5.   Aserinsky, E. and N. Kleitman. Regularly occurring periods of eye motility and concomitant phenomena during sleep. Science, 1953, 118:273–4.

6.   Brown, C. The stubborn scientist who unraveled a mystery of the night. Smithsonian magazine, October 2003, pp. 92–97.

7.   Dement, W. C. and N. Kleitman. The relation of eye movements during sleep to dream activity: An objective method for the study of dreaming. Journal of Experimental Psychology, 1957, 53(3):339–46.

8. Whitman, R. M., M. Kramer, P. H. Ornstein, and B. J. Baldridge. The Physiology, Psychology, and Utilization of Dreams. American Journal of Psychiatry, 1967, 124:43–58.

9. Hartley, D. *Observations on Man, His Frame, His Duty, and His Expectations*. London: Johnson, 1802.

Chapter 6, "Ed Evarts and Mickey Mantle"

1. Ed Evarts and Mickey Mantle Dream, JAH Journal Vol. 23, 7/5/1986.

2. Dement, W. The effect of dream deprivation. Science, 1960 Jun 10; 131:1705–7.

3. Hobson, J. A., R. W. McCarley, R. T. Pivik, and R. Freedman (1974). Selective firing by cat pontine brain stem neurons in desynchronized sleep. Journal of Neurophysiology, 37:497–511.

4. Hobson, J. A., R. W. McCarley, R. Freedman, and R. T. Pivik (1974). Time course of discharge rate changes by cat pontine brain stem neurons during the sleep cycle. Journal of Neurophysiology, 37:1297–1309.

5. McCarley, R. W. and J. A. Hobson (1975). Discharge patterns of cat pontine brain stem neurons during desynchronized sleep. Journal of Neurophysiology, 38:751–766.

6. Hubel, D. H. Single unit activity in striate cortex of unrestrained cats. Journal of Physiology, 1959 Sep 2; 147:226–38.

7. Pittenger, C. and E. Kandel. A genetic switch for long-term memory. Comptes Rendues dell'Academie des Sciences, 1998 Feb–Mar; 321(2–3):91–6.

8. Pavlov, I. P. *Conditioned Reflexes: An Investigation of the Physiological Activity of the Cerebral Cortex*. New York, Dover Publications, 1960.

9. Skinner, B. F. *The Behavior of Organisms: An Experimental Analysis*. New York: Appleton-Century-Crofts, 1966.

10. Stickgold, R., J. A. Hobson, R. Fosse, and M. Fosse (2001). Sleep, learning and dreams: Off-line memory reprocessing. Science, 294, 1052–1057.

11. Buzsaki, G. Memory consolidation during sleep: a neurophysiological perspective. Journal of Sleep Research, 1998; 7 Suppl 1:17–23.

12. Edelman, Gerald M. *Bright Air, Brilliant Fire: On the Matter of the Mind*. New York: Basic Books, 1992.

Chapter 7, "Bicentennial Wine Tasting"

1. Bicentennial Wine Tasting Dream, JAH Journal Vol. 26, 3/11/1987.

2. Nielsen, T. A. A review of mentation in REM and NREM sleep: "covert" REM sleep as a possible reconciliation of two opposing models. Behavioral and Brain Sciences, 2000 Dec; 23(6):851–66.

3. Maquet, P., J. M. Peters, J. Aerts, G. DelFiore, C. Degueldre, A. Luxen, and G. Franck. Functional neuroanatomy of human rapid-eye movement sleep and dreaming. Nature, 1996, 383(6596):163–66.

4. Braun, A. R., T. J. Balkin, N. J. Wesenten, R. E. Carson, M. Varga, P. Baldwin, S. Selbie, G. Belenky, and P. Herscovitch. Regional cerebral blood flow throughout the sleep-wake cycle. An H2(15)O PET study. Brain, 1997 Jul; 120 (Pt 7):1173–97.

5. Nofzinger, E. A., M. A. Mintun, M. Wiseman, D. J. Kupfer, and R. Y. Moore. Forebrain activation in REM sleep: an FDG PET study. Brain Research, 1997 Oct 3; 770(1–2):192–201.

6. Pinker, S. *The Language Instinct: How the Mind Creates Language*. New York: Perennial, 2000.

7. Gross, M. et al. Sleep Disturbances and Hallucination in the Acute Alcoholic Psychoses, unpublished manuscript, author's collection.

Chapter 8, "Dangerous Diving"

1. Dangerous Diving Dream, JAH Journal Vol. 29, 5/19/87.

2. Etherington-Smith, Meredith. *The Persistence of Memory: A Biography of Dali.* New York: Random House, 1992.

3. Hobson, J. A. and R. W. McCarley (1977). The brain as a dream state generator: an activation-synthesis hypothesis of the dream process. American Journal of Psychiatry, 134(12):1335–1348.

4. Hobson, J. A. Sleep and dream suppression following a lateral medullary infarct: a first-person account. Consciousness and Cognition, 11(3):377–90.

5. Lorente de Nò, R. Vestibulo-Ocular Reflex. Archives of Neurology and Psychiatry, 1933 30:245–91.

6. Bowker, R. M. and A. R. Morrison. The startle reflex and PGO spikes. Brain Research, 1976 Jan 30; 102(1):185–90.

7. Domhoff, G. W. The scientific study of dreams: neural networks, cognitive development, and content analysis. Washington, DC: American Psychological Association, 2003.

Chapter 9, "Split Whale"

1. Split Whale Dream, JAH Journal Vol. 23, 2/10/89.

2. Baghdoyan, H. A., A. P. Monaco, M. L. Rodrigo-Angulo, F. Assens, R. W. McCarley, and J. A. Hobson (1984). Microinjection of neostigmine into the pontine reticular formation of cats enhances desynchronized sleep signs. Journal of Pharmacology in Experimental Therapy, 231:173–180.

3. Baghdoyan, H. A., M. L. Rodrigo-Angulo, R. W. McCarley, and J. A. Hobson (1987). A neuroanatomical gradient in the pontine tegmentum for the cholinoceptive induction of desynchronized sleep signs. Brain Research, 414:245–261.

4.   Moldofsky, H. The contribution of sleep medicine to the assessment of the tired patient. Canadian Journal of Psychiatry, 2000 Nov; 45(9):798–802.

5.   Gazzaniga, M. and K. Baynes. Consciousness, Introspection, and the Split Brain: The Two Minds/One Body Problem. In: *The New Cognitive Neurosciences*, Second Edition, ed. M. S. Gazzaniga. Cambridge: MIT Press, 2000.

6.   Davidson, R. J. Affective neuroscience and psychophysi-ology: toward a synthesis. Psychophysiology, 2003 Sep; 40(5):655–65.

Chapter 10, "Tiffany Box"

1.   Tiffany Box Dream, JAH Journal Vol. 23, 10/18/1985.

2.   Armstrong, D. M. *A Materialist Theory of the Mind*. New York: Humanities Press, 1968.

3.   Grunbaum, A. *The Foundations of Psychoanalysis: A Philosophical Critique*. Berkeley: University of California Press, c1984.

4.   Wundt, W. *Grundzuge der Physiologische Physchologie*. Leipzig: W. Engelman, 1874.

5.   Fosse, R., R. Stickgold, and J. A. Hobson (2001). Brain-mind states: Reciprocal variation in thoughts and hallucinations. Psychological Science, 12, 30–36.

6.   Chalmers, D. J. *The Conscious Mind: In Search of a Fundamental Theory*. New York: Oxford University Press, 1996.

7.   Flanagan, Owen J. *The Problem of the Soul: Two Visions of Mind and How to Reconcile Them*. New York: Basic Books, 2002.

8.   Lavie, P. and J. A. Hobson (1986). Origin of dreams: Anticipation of modern theories of the philosophy and physio-logy of the eighteenth and nineteenth centuries. Psychiatric Bulletin, 100:229–240.

Chapter 11, "Louis Kane Dies"

1. Louis Kane Dream, JAH Journal 7/11/2002.

2. Stickgold, R., J. A. Hobson, R. Fosse, and M. Fosse (2001). Sleep, learning and dreams: Off-line memory reprocessing. Science, 294, 1052–1057.

3. Williams, J., J. Merritt, C. Rittenhouse, and J. A. Hobson (1992). Bizarreness in dreams and fantasies: Implications for the activation-synthesis hypothesis. Consciousness and Cognition, 1:172–185.

4. Cartwright, R. D., H. M. Kravitz, C. I. Eastman, and E. Wood. REM latency and the recovery from depression: getting over divorce. American Journal of Psychiatry, 1991 Nov; 148(11):1530–5.

Chapter 12, "Medieval Town"

1. Medieval Town Dream, JAH Journal Vol. 115a, 3/12/2001.

2. Hobson, J. A. Sleep and dream suppression following a lateral medullary infarct: a first-person account. Consciousness and Cognition, 11(3):377–90.

3. Hobson, J. A. Shock Waves: A Scientist Studies His Stroke. Cerebrum, 4(2): 39–57.

4. Hobson, J. A. The Enduring Self: A First Person Account of Brain Insult Survival. In: *The Lost Self: Pathologies of the Brain and Identity*, ed. F. Stevens. Cambridge: Oxford University Press, in press.

5. Antrobus, J. S. Dreaming: Cognitive processes during cortical activation and high afferent thresholds. Psychological Reviews, 1991, 98:96–121.

6. Stickgold, R., E. Pace-Schott, and J. A. Hobson (1994). A new paradigm for dream research: Mentation reports following spontaneous arousal from REM and NREM sleep recorded in a home setting. Consciousness and Cognition, 3:16–29.

7.   Werner D. L., K. J. Ciuffreda, and B. Tannen. Wallenberg's Syndrome: a first-person account. Journal of the American Optometric Association, 1989, 60(10): 745–747.

Chapter 13, "French Kiss"

1.   French Kiss Dream, JAH Journal 11/3/2003.

2.   Luppi, P. H. The Paradox of Sleep: A Symposium in Honor of Michel Jouvet. Archives Italiennes de Biologie, in press.

Epilogue

1.   Moruzzi, G. and H. W. Magoun. Brainstem reticular formation and activation of the EEG. Electroencephalography and Clinical Neurophysiology, 1949, 1:455–73.

2.   Aserinsky, E. and N. Kleitman. Regularly occurring periods of eye motility and concomitant phenomena during sleep. Science, 1953, 118:273–4.

3.   Berger, H. Uber das Elektrencephalogram des Menchen. Zweite Mitteilung. Journal of Neurology and Psychology, 1930, 40:160–179.

4.   Freud, S. The Future of an Illusion. New York: H. Liveright, 1928.

5.   Loomis, A. L., E. N. Harvey, and G. A. Hobart. Cerebral States During Sleep as Studied by Human Brain Potentials. Journal of Experimental Psychology, 1937, 21:127–44.

6.   Jouvet, M. and F. Michel. Correlation electromyographiques du sommeil chez le chat decortique mesencephalique chronique. Comptes Rendues des Seances de la Societe de Biologie et de Ses Filiales, 1959, 153:422–25.

7.   Pompeiano, O. Cholinergic Activation of Reticular and Vestibular Mechanisms Controlling Posture and Eye Movements. In: *The Reticular Formation Revisited*, eds. J. A. Hobson and M. A. B. Brazier. New York: Raven Press, 1979, pp. 473–572.

8.   Hobson, J. A., R. W. McCarley, and P. W. Wyzinski (1975). Sleep cycle oscillation: reciprocal discharge by two brain stem neuronal groups. Science, 189:55–58.

Acknowledgments

I am grateful to Sigmund Freud for creating a theory of dreaming that was clear enough to refute.

Most of the people who trained me and collaborated with me in my scientific research are mentioned in the text. One of them is the late Elwood Henneman, a classical Sherringtonian if there ever was one, but also a gentleman and a scholar of the old school. It is a pleasure to dedicate this book to him. Elwood took me in when I was scientifically homeless after my stay in Lyon, and it was in his lab that I made my first successful microelectrode recordings.

Many teachers prepared my mind for this project. At Harvard Medical School, I met the late Mark D. Altschuele who imbued me with his personal passion for physiology and his deep distrust for psychoanalysis. Mark helped me shape my own agenda through his generous support of our informal residents' seminar. That colloquium provided the opportunity to exchange critical ideas with Earnest Hartman, Anton Kris, Eric Kandel, John Merrifield, Stanley Palombo, George Vaillant, and Paul Winder. I thank them all, but especially Mark.

My scientific work was supported by several public and private agencies, chief among them the NIMH and the MacArthur Foundation.

Several people helped this book to be born. There was no more important literary midwife than Stephen Morrow, who realized that dream theory was a good jumping-off place for discussion of a new, brain-based psychology. Stephen also thought up the title of the book and helped me with the chapter subtitles. Nick Tranquillo changed my hen-scratch drafts into electronic form with great skill and patience. I am also thankful for the forbearance and affectionate support of my wife, Lia Silvestri, and the four children who lived with us in Sicily and in Boston while *13 Dreams* was being written.

Index